Cyanobacteria

NIPA® GENX ELECTRONIC RESOURCES & SOLUTIONS P. LTD.
New Delhi-110 034

Cyanobacteria

Antibacterial Activity

PURSHOTAM KAUSHIK, M.Sc., Ph.D.
Professor, Department of Botany and Microbiology
Gurukul Kangri University, Hardwar-249 404 (Uttarakhand)

ABHISHEK CHAUHAN, M.Sc. Ph.D. (Microbiology)
Senior Microbiologist
Shriram Institute of Industrial Research, New Delhi

NIPA® GENX ELECTRONIC RESOURCES & SOLUTIONS P. LTD.
New Delhi-110 034

NIPA® GENX ELECTRONIC RESOURCES & SOLUTIONS P. LTD.

101,103, Vikas Surya Plaza, CU Block
L.S.C. Market, Pitam Pura, New Delhi-110 034
Ph : +91 11 27341616, 27341717, 27341718
E-mail: newindiapublishingagency@gmail.com
www: www.nipabooks.com

For customer assistance, please contact
Phone: + 91-11-27 34 17 17
Fax: + 91-11- 27 34 16 16
E-Mail: feedbacks@nipabooks.com

ISBN: 978-81-19072-77-4

Composed and Designed by NIPA®.

Dedicated to

Dr. Satish Chander Verma

Dr. Satish Chander Verma (b. 19 December, 1931), Professor of Botany, FNASc, FLS, FBS, FSCG, FIFS, and former Head of the Department of Botany at Panjab University, Chandigarh, is known for his contributions in cytology, cytogenetics, genetics, and reproductive and evolutionary biology of homosporous ferns and *Isoetes*. Besides, his contribution to the genetics of self-incompatibility in Brassicaceae, nuclear DNA estimations in species of *Brassica* of U's triangle, Giemsa C-banding studies in Rye, and *Lathyrus*, are exemplary. He has the distinction of working with Professor Hugh Rees, FRS, and Professor Dan Lewis, FRS, in UK, supported by British Council. Professor Verma has vast experience of teaching and research (since 1956), and has guided a large number of students for their M. Phil. and Ph.D. degrees, and has over 200 publications to his credit in national and international journals. Dr. Verma participated in several national and international conferences and symposia. He has been awarded Royal Society, London, Bursary (UK); Visitor-ship of British Council (UK) at University College (London) & Queen Mary College (London), and Visiting Professorship at Hiroshima, Tokyo, and Sendai (Japan), University College, London (UK), Colorado College (USA), and University of Arizona (Tempe, USA), Honorary Research Fellow of University College, London(1977-1984), UGC National Lecturer (1986-87), UGC Emeritus Fellow (1995 &1996), PN Mehra Memorial Medal, and DD Pant Oration Medal.

The lord created medicines out of the earth,
and the sensible will not despise them.

Selman A. Waksman

The legend goes on that the medicine should always be taken under the supervision of a qualified physician.The cyanobacteria were believed to be toxic in the past which is true even today nonetheless they have been reported to be of therapeutic value in the literature. Here, lies the potential probability to exploite them as alternative to antibiotics for pharmaceutical purposes after more intensive trials.

An Intuitive Opinion

Preface

Emergence of new pathogens and increasing antibiotic resistance among existing microorganisms has become a major problem to the humanity. The time has come to get rid of this problem and the biologists should rise to the occasion to discover novel sources of medicine.

Taking this into consideration, a global attention has arisen towards finding new chemicals, with lead structures, which either directly or after small modifications can be used for development of new drugs.

The advantage of microorganisms, which can be cultured, such as cyanobacteria, is that a sustainable supply of desired metabolites can be achieved, which is not always possible for a product derived from sources based on macro-organisms. Moreover, cyanobacteria promise greater opportunities with all relatively inexpensive and easier approaches against some expensive and technically complex methods presently in use for new drug discovery, such as - combinatorial chemistry and molecular biotechnology.

The various properties of cyanobacteria include antibacterial, antifungal, anti-HIV, anticancerous and antialgal characteristics.

Present study deals with some cyanobacterial species from Nostocaceae, Oscillatoriaceae and Stigonemataceae. Each culture was harvested and spent medium was collected by filtration of the biomass. Hexane, ethyl acetate, dichloromethane and methanol were used as the solvents for obtaining the extracts from biomass of each culture. The extracts and spent medium thus obtained were investigated for antibacterial activity against selected bacterial strains.

This research monograph is a novel attempt of this type and will be useful for the research students and teachers of the Universities and colleges who want to study microbiology, medicine and pharmaceutical sciences with respect to antibacterial effects. It will also be useful for the pharmaceutical industries, for research and development (R&D) related to drug discovery.

Prof. Purshotam Kaushik
Dr. Abhishek Chauhan
Department of Botany and Microbiology
Gurukul Kangri University, Hardwar
e-mail : purshotam.kaushik@gmail.com
purshotam.kaushikrediffmail.com
website : purshotam.kaushik.googlepages.com

Acknowledgement

Prof. B.D.Kaushik and Dr. Rajesh Kumar
Department of Microbiology
Indian Agricultural Research Institute, New Delhi

Dr. Mukund Sharma
Birbal Sahni Institue of Palaeobotany
53, University Road, Lucknow, U.P.

Prof. A. Bohra and Dr. Anil Vyas
Department of Botany, JNV University, Jodhpur

Dr. A. K. Dhiman
Information Scientist
Gurukul Kangri University, Hardwar

Mr. Pankaj Goyal, Sachin Tyagi
Arun Karanwal, Ms. Garima and Ms. Pallvi Pal
Gurukul Kangri University, Hardwar

Abbreviations

%	Percentage
µg	Microgram
µg/ml	Microgram per millilitre
^{0}C	Degree centigrade
CAMHB	Cation adjusted Muller Hinton broth
cfu	Colony forming unit
DCM	Dichloromethane
DNA	Deoxyribose nucleic acid
E	Ethyl acetate
e.g.	For example
et al.	et alii
Fig	Figure
g/ml	Gram per millilitre
GC-MS	Gas chromatography-mass spectroscopy
H	Hexane
HPLC	High performance liquid chromatography
i.e.	That is
IR	Infrared
LC-MS	Liquid chromatography-mass spectroscopy
M	Methanolic extract
mg	Milligram

mg/ml	Milligram per millilitre
MHA	Muller Hinton agar
MHB	Muller Hinton broth
MIC	Minimum inhibitory concentration
ml	Millilitre
mm	Millimeter
MS	Mass spectroscopy
MTCC	Microbial type culture collection
NaCl	Sodium chloride
NCCLS	National committee for clinical laboratory Standards
nm	Nanometer
NMR	Nuclear magnetic resonance
O.D.	Optical density
pH	Hydrogen ion concentration
ppm	Parts per million
rpm	Rotation per minute
sp.	Species
TLC	Thin layer chromatography
UV	Ultra violet
viz.	Namely

❑❑❑

Contents

Dedicated to Dr. Satish Chander Verma v
Citation of Waksman and An Intuitive Opinion vi
Preface vii
Acknowledgement ix
Abbreviations xi

Chapter 1 : Introduction ____ 1
Chapter 2 : Review of Literature ____ 21
Chapter 3 : Bacteriology of Test Microorganisms ____ 45
Chapter 4 : Material and Methods ____ 81
Chapter 5 : Results ____ 97
Chapter 6 : Discussion ____ 137
Chapter 7 : Summary ____ 145
Chapter 8 : Conclusion ____ 151
Chapter 9 : Literature Cited ____ 153
Chapter 10 : Glossary ____ 173

Index 193

Chapter 1

INTRODUCTION

Some cyanobacteria (Blue-green algae) like *Nostoc* are known to work as one of the two coponenents (photosynthetic partner or phycoboint) of lichens. The second partner of lichen is a lichen fungus or mycobiont. The lichens have not been described as a therapeutic agent in Charak Samhita, the monumental work on *materia medica* prepared by great Acharya Charak and his group in India more than three thousand years ago. However, these have been used in folklore medicine in many countries of the world and have been regarded to bear antibacterial properties (Katiyar and Katiyar, 2009). They are found in tundra even, the treeless plain of the arctic regions of Russia, Siberia, marshy in summer and frozen hard in winter. It is common to see gunny bags full of lichens for sale in the Masala market near Kesa House, Express Canal Road, at Kanpur. They also comprise a part of meat masala (spices for meat). There are references that Raja (King) Todarmal studied algae in relation to purification of water.

However, the search for antibiotics began in the late 1800s, with the growing acceptance of the germ theory of diseases; a theory which linked bacteria and other microbes to the causation of a variety of ailments. As a result, scientists began to devote time searching for drugs that would kill these disease-causing bacteria. The goal of such research was to find so called 'Magic Bullets or wonder drugs' that would destroy microbes without toxicity to the person taking that drug.

Today, most of the diseases caused by pathogens can be cured with the help of available antibiotics, but discovery of any new antibiotic, generally follow up with a course of resistance mechanism building up against it among the target organisms. This phenomenon known as -'antibiotic resistence' is developing among microbial species at an appreciable rate, is a formidable complication of prudent and overuse of available antibiotics and is imposing a serious health-threat to human welfare.

There are unfortunately, frequent instances of multiple resistance to antibiotics as well. Thus the bacteria with this multiple resistance are extremely hazardous and threaten to put the clock back to times before the introduction of antibiotics. Consequently, there arises a need to explore and develop new effective antibiotics against microbial pathogens. Taking this into consideration, there has been a global attention towards finding new chemicals, which led to development of structures, which either directly or after small modifications can be used for development of new drugs. This prompted researchers to search for new sources of antibiotics.

The advantage of microorganisms, which can be cultured, such as cyanobacteria, is that a sustainable supply of desired metabolites can be achieved, which is not always possible for a product derived from a macro-organism source, besides cyanobacteria promise greater opportunities with all relatively inexpensive and easier approaches against some expensive and technically complex methods presently in use for new drug discovery, such as - combinatorial chemistry and molecular biotechnology.

Screening of cyanobacteria for antibiotics and pharmaceutically active compounds has received ever increasing attention. The molecules isolated from several cyanobacteria have broad spectrum of biological activities including toxins, antibiotics, fungicides and algicides (Browitzka, 1995). Random screening of cyanobacteria will continue to play an important role in the drug discovery process for the foreseeable future. Historically, actinomycetes have been the most prolific producers of metabolites with significant antimicrobial activities. They also have been, not surprisingly, the most closely examined group of organisms and at the present time, yield known compounds at a rate in excess of 95% of all active leads discovered in primary screening (Bradner, 1980).

The cultivable microorganisms for screening are thus of paramount importance. The search for cyanobacteria with antimicrobial activity has gained importance in recent years due to growing world wide concern about alarming increase in the rate of infection by antibiotic-resistant pathogen and newly emerged diseases (Hayashi *et al.*, 1994). They can be cultured in the laboratory with relative ease to provide a consistent source of biologically active secondary metabolites (Patterson *et al.*, 1993).

The cyanobacteria are the group of structurally diverse Gram-negative eubacteria characterized by their ability to perform oxygenic photosynthesis. They are the prokaryotic microscopic organisms which do not have true nucleus and double-membranous structure like mitochondria, chloroplast etc. Cell diameter ranges from less than 1.0 mm to over 100 mm. The really interesting thing about some cyanobacteria is that they can almost be considered multicellular organisms, because some cells undergo cell differentiation. Their cells exist in colonies of long straight or branching chains, called filaments or trichomes (Holt *et al.*, 1994). DNA lies freely in the center of the cell known as centroplasm. Their structural diversity is parallel to bacteria by their genetic heterogeneity, reflected in G+C values (35% to 71%) and genome size (M.W. of 1.6×10^6 to 8.6×10^9 daltons), that span nearly the entire range found among bacteria.

The cyanobacteria occur in almost all conceivable habitats. Most species of cyanobacteria are free-living, freshwater, marine or terrestrial: planktonic, or benthic; and comprise major components of microbial mats. Some are thermophilic, growing in hot springs. A few cyanobacteria are symbionts of liverworts, water ferns, and cycads; while a number of them are found as the phototrophic component of lichens. In the case of the water fern *Azolla*, it has been shown that the cyanobacterial endophyte fixes up nitrogen that becomes available to the plants (Madigan *et al.*, 2000; Kaushik, 2000; Kaushik, 2004, 2008).

They are the most efficient converters of solar energy to useful forms so they led to increased interest in techniques for their isolation and purification (Stanier, 1988). Cyanobacteria do not require carbon or energy sources in their growth medium. Thus, they require only basic inorganic medium, which has several logical advantages when performing the mass culture and purification of active compounds.

According to Holt *et al.* (1995) light harvesting pigments of cyanobacteria includes phycobiliproteins or just phycobilins which are phycoerythrin (PE, in some), phycoerythrocyanin (PEC in some), phycocyanin (PC in all) and allophycocyanin (APC in all), chlorophyll a (Chl a) and carotenoids (especially β- carotene, zeaxanthen, echimenone, myxoxanthophyll and oscilloxanthin). PE may contain only phycoerythrobilin as chromophore (C-PE) or a mixture of phycoerythrobilin and phycourobilin. All known species of cyanobacteria have some PC and APC. Growth rate and production of metabolites in cyanobacteria is influenced by cultural conditions, optimization of media, time of harvest and physical parameters such as light and temperature, and then they can lead to greatly enhanced product yield. The generation time for cyanobacteria is relatively slow and typical doubling time ranging from 0.3 to 1.4/day under optimum conditions (Kaushik, 1987).

The cells of cyanobacteria are surrounded by envelopes consisting of the cytoplasmic membrane and cell wall. Cell wall is four layered and consists of murein, outside to which lipopolysaccharide layer is present. The presence of an outer membrane indicates a Gram-negative cell wall organization. The peptidoglycan of cyanobacteria has several properties in common with that of Gram-positive bacteria, such as thickness, degree of cross-linkage, and presence of covalently linked polysaccharides. Furthermore, in contrast to capsules of bacteria, sheath of cyanobacteria are often directly visible in the light microscope (Weckesser and Jurgens, 1988; Kaushik, 2004, 2008).

Generally all cyanobacterial vegetative cells contain carboxysomes, pseudocrystalline aggregates of the key enzymes of CO_2 fixation *via* the reductive pentose phosphate pathway and glycogen is a general carbohydrate reserve material of cyanobacteria. Other cellular inclusions include Poly-β-hydroxybutyrate (PHB) granules, cyanophycean granules, polyphosphate granules, carboxysomes or polyhedral bodies and gas vesicles (Stanier, 1988).

It has been accepted in general that Light-harvesting pigments include chlorophyll-a, phycobilliproteins, phycocyanin, phycoerythrin in some, and carotenoids (especially β-carotene and zeaxanthin). Because of the chlorophyll-a, the cells absorb red light of 680 nm. The photosynthetic apparatus is present in the form of thylakoids, which are flattened membranous sacs located within the cell. The surface of thylakoids is studded with phycobilisomes, which contain phycobilin pigment. However, Chlorophyll-b and chlorophyll-c are absent (Holt *et al.*, 1994; Pelczar *et al.*, 1993).

It is interesting to note that the blue green algae are among the oldest photoautotrophic organisms; their cultivation without organic substrates can be an economical advantage over other microorganisms. Therefore, an optimized production of relevant compounds under controlled condition is conceivable (Kulik, 1995).

The cyanobacteria are known to produce a wide variety of toxins and other bioactive compounds, which include 40% lipopeptides, 5.6% aminoacids, 4.2% fatty acids, 4.2% macrolides and 9% amides (Fig.1).

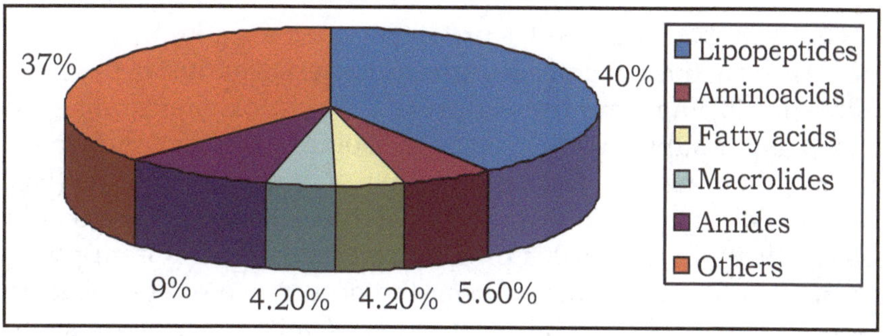

Fig. 1. Types of chemical compounds isolated from marine cyanobacteria (Burja *et al.*, 2001)

According to Burja *et al.* (2001) and Singh *et al.* (2001) the cyanobacterial lipopeptides include different compounds like cytotoxic (41%), antitumor (13%), antiviral (4%), antibiotics (12%) and 18% activity includes antimalarial, antimycotics, multi-drug resistance reversers, antifeedant, herbicides and immunosuppressive agents (Fig. 2).

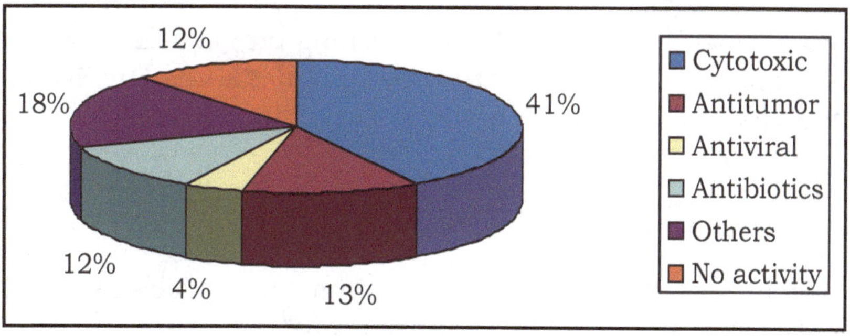

Fig. 2. Reported biological activities of cyanobacterial compounds (Burja *et al.*, 2001)

It has been estimated that about 2000 strains of cyanobacteria are distributed all over the world. They show a remarkable ecological diversity. Because of widespread eutrophication of lakes, ponds and some parts of oceans cyanobacteria often form blooms, which lead to water hygienic problems (Henning and Kohl, 1981; Skulberg *et al.*, 1984; Chorus and Bartram, 1999; Duy *et al.*, 2000). They may cause unpleasant tastes and odours through excretion of volatile compounds (Jones and Korth, 1995). Several genera of cyanobacteria form toxic water blooms and different cyanobacterial toxins have been characterized (Carmichael, 1992; Rinehart *et al.*, 1994; Hunter 1995).

The cyanobacteria besides producing toxins are known to have a wide array of active substances with antibacterial, antiviral, fungicidal, enzyme inhibiting, immunosuppressive, cytotoxic, and algaecidal activity. Such substances have been isolated from cyanobacterial biomass or in some cases from the medium of laboratory cultures (Falch, 1996; Hayashi *et al.*, 1996; Namikoshi and Rinehart, 1996; Banker and Carmeli, 1998). In the 1970, a pronounced reduction of Gram-positive bacteria was found in lakes during the occurrence of cyanobacterial blooms (Chrost, 1975). The production of antibacterial substances could be the reason of this phenomenon.

The cyanobacteria bear the characteristics to secrete vitamins, amino acid, fatty acid, carbohydrates and metabolites like primary and secondary amines, histamines, histidine, tannins, terpenoids, bromophenol and polysaccharides. Few of these compounds are proven to be biologically active (Metting *et al.*, 1986). The recent examples are cyanovirin-N secreted by *Nostoc elliposospoporum* and anti-HIV glycolipids secreted by *Isochyrosis* and bromophenol are secreted by *Calothrix sp.* (Jaspars and Lawton 1998).

They have become targets for screening programme in search of novel compounds of potential medicinal value. The study of cyanobacteria as a source of new chemical entities is an exciting field which needs to be investigated further.

Cyanobacteria and Earth History

The history of the Earth is written on rocks which can be read from *'bioscripts'* preserved by nature in the form of fossils. This history for the sake of convenience has been divided by geologists into four great eras :

1. **Pre Cambrian era** which started about 4500 million years age and ended 500 million years ago and is characterized by scarcity of fossils either of animals or plants.

2. **Palaeozoic era** characterized by marine invertebrates, marine fishes and ambhibians (Include *periods* or *systems* Cambrian, Ordovician, Silurian, Upper Silurian, Lower Debonian, Lower Carboniferous, Upper Carboniferous or Coal Measures, Lower Permian and Upper Permian).
3. **Mesozoic** Characterized by reptiles and ammonites.
4. **Cainozoic** extending to present day and characterized by land mammals.

Herrero and Flores, (2008) opined that the geological record indicates that the use of water as the source for electrons in photosynthesis evolved in a common ancestor of extinct cyanobacteria. An apparent study on cyanobacterial evolution reveals that the existence of blue-greens was even about 2000 Ma (Million years ago). Cyanobacteria remained principal primary producers throughout the Proterozoic Era (2500-543 Ma), in past because the redox structure of the oceans favoured photautotrophs capable of nitrogen fixation. Cyanobacteria remain critical to marine ecosystems as primary producers in oceanic gyres, as agents of biological nitrogen fixation, and, in modified form, as the plastids of marine algae.

The cyanobacteria have also been tremendously important in shaping the course of evolution and ecological change throughout earth's history. The oxygen atmosphere that we depend on was generated by numerous cyanobacteria photosynthesizing during the Archaean and Proterozoic Era. Before that time, the atmosphere had a very different chemistry, unsuitable for life as we know it today.

Fossil Record of Cyanobacteria

The oldest known fossils, in fact, are cyanobacteria from Archaean rocks of Western Australia, that have been detected to be 3.5 billion years old. This may be somewhat surprising, since the oldest rocks are hardly a little older *i.e.* 3.8 billion years old. Many Proterozoic oil deposits are attributed to the activity of cyanobacteria, such as *Gloeocapsomorpha*. Small concentrically layered structures called *pisolites* which are the sedimentary rocks formed from pisoliths are also the result of fossilized bacteria. Cyanobacteria are otherwise rarely preserved in rocks other than chert, though some possible blue-green bacteria have been recovered from shale or the soft rock that splits up easily (Taylor, 1993).

They are among the easiest microfossils to recognize. Morphologies in the group have remained much the same for billions of years, and they may leave chemical fossils behind as well, in the form of breakdown products from pigments. Small fossilized cyanobacteria have been extracted from Precambrian rock, and studied through the use of SEM and TEM (scanning and transmission electron microscopy).

The layered *stromatolites* are one of the examples produced by the activity of earliest Cyanobacteria. The oldest stromatolites date to the Early Archaean, and they became abundant by the end of the Archaean. In the Proterozoic, stromatolites were widespread on earth, and were ecologically important as the first reefs. By the close of the Proterozoic, the abundance of stromatolites decreased markedly, though cyanobacteria continued to leave a fossil record, such as *Langiella* and *Kidstoniella* known from the Lower Devonian Rhynie chert.

The *chert deposits* at Rhynie, about 8 feat thick are believed to represent a *peat bog* which got infiltrated with silica. Thus plant remains of upper Silurian and Lower Devonian rocks of Scotland of Britan became preserved. Leafless and rootless *Rhynia gwynne - vaughnii* was first of all described by Kidston and Lang in 1917 which bears ancestory to vascular cryptogams.

Two species of fossil cyanobacteria *Archaeothrix* were initially described by Kidston and Lang in 1921. These are *Archaeothrix contexta* and *Archaeothrix oscillatori formis*, both of which consist of simple unbranched filaments bearing *discoid cells* and *heterocysts*. These two species have been differentiated mainly on the basis of diameter of cells. The cells in *A. oscillatori formis* are broader than in *A. contexta.* Out of these two species *A. oscillatoriformis* was found inside the partially decayed stems of fossil pteridophyte *Rhynia gwynne - vaughanii.*

The serial efforts of Kingston and Lang in 1921, Croft and George in 1959 and Edwords and Lyon in 1983 resulted in describing and naming fossilized cyanobacteria *Archaeotrix contexta, A. oscillatori formis, Kidstoniella fristchii, Langiella scourfeldii, Rhyniella vermiformis* and *Rhyniococcus uniformis*.

These layers were produced as calcium carbonate precipitated over the growing mat of bacterial filaments; photosynthesis in the bacteria depleted carbon dioxide in the surrounding water, initiating the precipitation. The minerals, along with grains of sediment precipitating from the water, were trapped within the sticky layer of mucilage that surrounds the bacterial colonies, which continued to grow upwards through the sediment to form a new layer. As this process occurred over and over again, the layers of sediments were created. It would be interesting to note that this phenomenon of nature is yet in action. Shark Bay in Western Australia is well known for the *stromatolite* the surface soils or "turfs" rising along its beaches.

Fossilied Cyanobacteria from India

Prof. D. M. Banerjee, Indian National Science Academy, Senior Scientist of Department of Geology, University of Delhi, was one of the pioneer workers on fossilized cyanobacteria. Comprehensive work has also been done in this field by Dr. Mukund Sharma of Birbal Sahni Institute of Palaeobotany, Lucknow and several other workers.

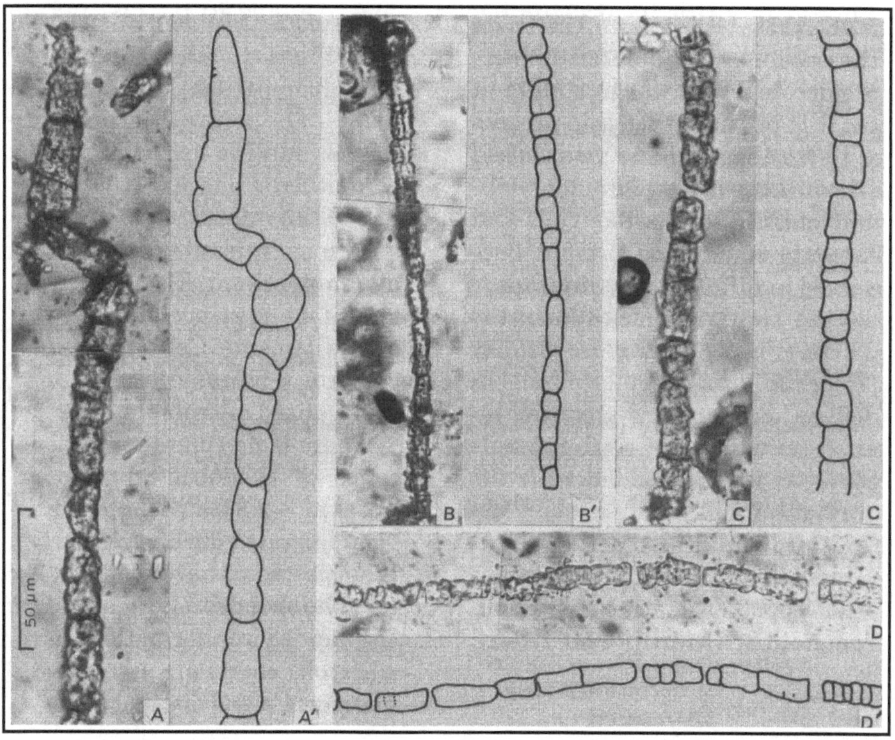

Fig. 3. Optical photomicrographs showing structurally preserved trichomes in thin section of black chert from the Donimalai formation of the Sandur Schist Belt, Dharwar, India. Scale is same all photographs = 50µm. Stage readings are for the Leitz Diaplan. Each photomicrograph is supported by a schematic representation of the fossil. A, A' (48.4; 102.3) and C, C' (53.0; 104.5)- *Phormidella sandurense*; B,B' (46.7; 97.5) and D,D' (47.8; 97.5) *Phormidella tenue*. Adapted from Venkatachala *et al.* (1990).

Cyanobacterial Evolution: Comparative Genomics

According to Herrero and Flores (2008), DNA sequencing offers a significant study on genetic mechanisms of adaptation. Similarly, comparative genomics-based approach towards understanding the evolution of Cyanobacteria is a great tool in current scientific industry. Basically, systematic methods of defining morphological traits in cyanobacteria have posed a major barrier in reconstructing their true evolutionary history. The initiation of protein sequencing, DNA sequencing and most notably the use of 16S rRNA as a molecular marker helped circumvent this barrier and now forms the basis of understanding the history of life on Earth. By integrating phylogenetic data inferred for hundreds to nearly 1000 protein coding genes common to all or most cyanobacteria, we are able to reconstruct an evolutionary history of the entire phylum, establishing a framework for resolving how their metabolic and phenotypic diversity came about.

Cyanobacteria as a Biopesticide

Dhananjaya *et al.* (2003) has reported that a freshwater cyanobacterium, *Microcystis aeruginosa*, and local soil isolate *Scytonema* sp. strain BT 23 were demonstrated to contain biotoxic secondary metabolites with pesticidal and mosquito larvicidal activities. A purified toxic constituent from *M. aeruginosa* showed an absorption maximum at 230 nm and its toxicity symptoms, Rf value on TLC, and retention time observed in an HPLC analysis were similar to those of the hepatotoxic heptapeptide microcystin-LR. The bioactive constituent of the *Scytonema* sp. was less polar in nature and exhibited two peaks at 240 and 285 nm. When applied to two cruciferous pests, *Pieris brassicae* and *Plutella xylostella*, the crude extracts and toxic principles from the two cyanobacteria showed significant antifeedant activity, and at higher concentrations exhibited contact toxicity to the insect larvae. The purified toxin from *M. aeruginosa* was found to be more effective and produced 97.5 and 92.8% larval mortality in the two pests, following 2 h of toxin treatment at a concentration of 25 μg per leaf disc (2.5 cm dia.). Meanwhile, similar treatment with the purified toxin from *Scytonema* sp. only produced 73 and 78% mortality in the two pests. The cyanobacterial constituents also showed significant activity against *Culex* and *Anopheles* larvae. The *M. aeruginosa* toxin (20 μg ml^{-1}) caused 98.2 and 88.1%, mortality in the *Culex* and *Anopheles* larvae, respectively, while the purified toxin from the *Scytonema* sp.

was less toxic and only produced a 96.3 and 91.2% mortality, respectively, at a much higher concentration (40 μg ml^{-1}).

According to Sathiyamoorthy and Shanmugasundaram (1996), a purified glycine-rich peptide toxin of cyanobacterium *Scytonema* MKU 106 showed an absorption maximum at 228 nm in UV spectra and the molecular mass was determined <12 kDa. The significant mortality rate was observed in case of American boll worms (*Helicoverpa armigera*) after treatment with crude and purified peptide toxins. The LC_{50} (lethal concentration to 50% of the population) for *Heliothis* larvae after 96 h was 8.37 g/ml purified peptide toxin and 6.27 g/ml crude peptide toxin. Observations also show that the peptide toxin at 0.01% concentration acts as a biopesticide and at high (0.1%) concentrations it will act as an anti-feeding compound for *Stylepta derogata* (leaf-roller) larvae of the cotton crop.

Cyanobacteria and Biosensor

Biosensor is a term used for a number of devices either used to monitor living systems or incorporating biotic elements. These sensors are primarily developed to detect analytes such as hydrogen peroxide (H_2O_2) which is of utmost importance in industrial (pharmaceutical, food, clinical) and environmental analyses. Its use as an antibacterial agent added to milk, and the environmental need to avoid halogenated substances for disinfection purposes, makes H_2O_2 an important substance in the food and beverage industry. This has raised extensive demands for establishing protocols for H_2O_2 detection depending on its application.

Compact amperometric algal biosensors for the evaluation of water toxicity : This sensor is developed using unicellular microalga *Chlorella vulgaris* (Chlorophyceae) which was entrapped in an alginate gel or a polyion complex membrane, immobilized directly on the surface of a transparent indium tin oxide electrode (Isao *et al.*, 2005). Photosynthetically generated oxygen of the immobilized algae was monitored amperometically. Responses of the algal biosensor to four toxic compounds, 6-chloro-N-ethyl-N-isopropyl-1,3,5-triazine-2,4-diamine (atrazine), 3-(3,4-dichlorophenyl)-1,1-diethylurea (DCMU), toluene and benzene, were evaluated as inhibition ratios of the reduction current. The concentrations that give 50% inhibition of the oxygen reduction current (IC_{50}) for atrazine, DCMU, toluene and benzene were 2.0, 0.05, 1550 and 3000 μmol dm^{-3}, respectively. There was a good correlation between these data and those of the conventional standard growth test. In comparison with the conventional algal biosensors based on the Clark-type oxygen electrode, the present sensor is much smaller and less expensive, and its assay time is much shorter(Shitanda *et. al.,* 2005).

Whole-cell biosensor for determination of volatile organic compounds in the form of aerosols : A biosensor using *Chlorella* microalgae immobilized on the membrane of an oxygen electrode has been designed to operate in the gas phase for determination of volatile organic compounds in the form of aerosols (Martine and Canh, 1998). Perchloroethylene (tetrachloroethylene) is used as an example to produce aerosols and vapours since it is toxic to the people working in dry-cleaning atmospheres and municipal treatment plants. Monitoring of perchloroethylene is achieved by the measurement of oxygen production during the algae photosynthetic process. The use of whole cells as bioreceptors allows the detection of a toxic aerosol in the gas phase without preliminary trapping of the analyte in a solution.

Cyanobacterial biofertilizer

Traditionally agriculture was dependent on natural inputs like farm yard manure, green manure or on microbial activity (Kaushik, 1994). However, with the introduction of inorganic fertilizer as well as fertilizer responsive varieties of various crops, organic inputs continue to decline. It resulted in decreased organic matter (humus) in soil which affected fertility of soil severely. For example, the organic carbon content of Punjab soils was 0.5% in 1980 whereas today it is only 0.2%. Hence serious attention was paid by scientists to exploit micoorganisms which are directly responsible to replenish the soil with certain nutrients like fixation of atmospheric nitrogen to ammonia, solubilization of phosphorus as well as addition of organic carbon. With the introduction of high yielding fertilizer responsive varieties of rice, demand for nitrogenous chemical fertilizers has increased several-fold. Therefore, it is essential to develop non-fossil fuel based nutrient alternatives with better utilization efficiency.

Nitrogen is an essential nutrient required for plant growth. It is present as diatomic elemental nitrogen and constitutes more than 79% of total gases. However, nitrogen in elemental form cannot be utilised by crop plants and crops starve amidst plenty. Nutrient nitrogen as such is the most limiting nutrient in soil and use of inorganic nitrogen fertilizers to meet the crop requirements is unavoidable for intensive farming practices. On global basis out of 180 million tonnes of nitrogen added annually on the earth's surface, the industrial fertilizer nitrogen accounts for only 1/3, while 2/3 comes from biological reduction of atmospheric nitrogen. The role of the diazotrophic reduction in sustainability of soil fertility is therefore, more important. The dinitrogen reduction so far is confined to prokaryotic cellular forms in nature. Among these prokaryotes, the photosynthetic diazotrophs - Cyanobacteria, are free living as well as symbionts with lichens, ferns and cycades and their contribution in total nitrogen fixation by biological means probably comes next to the legume-*Rhizobium* symbiosis. The cyanobacteria are ubiquitous in nature and are next to bacteria in distribution. They grow wherever water, sunlight and limited number of inorganic elements available and rice field ecosystem is the ideal niche for the growth and proliferation of cyanobacteria. Due to their wider occurrence in rice field ecosystem, these organisms have been exploited as a means to provide biologically fixed nitrogen.

Cytotoxic Activity of Cyanobacteria

Fish and Codd (2006) studied the cytotoxic activity of *Phormedium sp.* (a thermotolerant cyanobacterium). Falch *et al.* (1992) isolated polyhalogenated aromatic compound from *Fischerella ambigua* having cytotoxic activity. Cytotoxic antibiotics discovered during screening of some cyanobacteria are listed in table 1.

Table 1. Cytotoxic antibiotic discovered during screening of Cyanobacteria

Cyanobacteria	Antibiotic	References
Oscillatoria	Acutiphycins	Barchi *et al.*, 1984
Nostoc	Indolocarbazoles	Knubel *et al.*, 1990
Scytonema	Mirabilene Isonitriles	Carmeli *et al.*, 1990a
Cylindrospermum, Nostoc	Paracyclophanes	Moore *et al.*, 1990
Scytonema	Scytophycins	Ishibashi *et al.*, 1986; Carmeli *et al.*, 1990b
Scytonema	Tantazoles	Carmeli *et al.*, 1990c
Scytonema	Tolytoxin	Carmeli *et al.*, 1990b
Tolypothrix	Toyocamycin	Stewart *et al.*, 1988
Tolypothrix	Toyocamycin-5′-alpha -D-glucopyranose	Stewart *et al.*, 1988
Tolypothrix	Tubercidin-5′-alpha-D-glucopyranose	Stewart *et al.*, 1988
Plectonema	Tubercidin	Stewart *et al.*, 1988

Cyanobacteria (Biomodulatory Effect)

Biomodulatory effect of Cyanobacteria was reported by Singh *et al.*, 2005. Various species of cyanobacteria have biomodulatory effects. Oral doses of *Apbanizomenon flosaquae* on healthy human revealed a slight decrease in phagocytic activity of polymorph nucleated cells in *in vitro* conditions (Jensen *et al.*, 2000). This may indicate an anti-inflammatory rather than antiphagocytic effect on human neutrophils. On feeding the cyanobacterial species, the spleenic leucocytes from chickens showed greater anti-tumor activity (Qureshi *et al.*, 1996).

Effect of Cyanobacteria on specific Immunity : It was reported that by injecting the cyanobacterial extracts intraperitoneally 1 h prior to allergic challenge, mortality induced by the anaphylactic compound decrease (Yang *et al.*, 1997; Kim *et al.*, 1998). In another study, Hayashi *et al.* (1994) reported that immunized mice fed with a cyanobacterial supplement diet showed increased numbers of splenic IgM antibody-producing cells. Cyanobacteria do not seem to induce or enhance the food allergic IgE-dependent reaction. However, when ingested along with or before a potential antigen, they may enhance the IgA antibody level to protect against food allergies (Hayashi *et al.*, 1998).

Effect on Leucocyte Trafficking : Jensen *et al.* (2000) demonstrated that the cyanobacterium *Apbanizomenon flosaquae* triggers the migration of 40 % of natural killer cells within 2 h. *Apbanizomenon flosaquae* was also shown to stimulate the mobilization of T and B cells.

Anti-inflammatory Activity of Cyanobacteria : Cyanobacteria contain significant amount of carotenoids *i.e.* β-carotene, lycopene, lutein having antioxidant properties. By the quenching action on the reactive oxygen species, these carotenoids also have anti-inflammatory activity. The anti-inflammatory activity is also due to phycocyanin, a photo harvesting pigment. The anti-inflammatory effect seemed to be result of phycocyanin inhibiting the formation of leucotriene, an inflammatory metabolite of arachidonic acid (Romay *et al.*, 1999). *Apbanizomenon flosaquae* decrease the level of arachidonic acid (Kushak *et al.*, 2000). *Apbanizomenon flosaquae* and *Spirulina* contain significant amount of omega-3-alpha linolenic acid which inhibit the formation of inflammatory postaglandins and arachidonate metabolite.

Antihelmintic Activity

Dihydroxytetrahydrofuran from the south Australian marine brown algae *Notheia anomala* exhibits selective nematodal activity against *Trichostorgyrus* with a lethal dose in 50% of LD_{50} of 9.9 μg/ml. This inhibits the development of eggs for the infective free-living stage. The sponge geodin A is macrocyclic polyketide lactam tetramic acid, showing an LD_{50} value 1.0 μg/ml (Donia and Hamann, 2003). Jasplakinolide is a potent antiprasitic and antifungal agent and exhibits *in vitro* 50% effective dose of less than 1 μg/ml against nematode *Nippostrongylus braziliensis*.

Anticancer Activity of Cyanobacteria

Research by Jaspers and Lawton (1998) has focussed on various biologically active compounds from cyanobacteria. The *curian A,* a novel lipid compound isolated from *Lyngbya majuscula,* is a potent inhibitor of microtubule assembly with very low IC_{50} values against L1210 leukemia cells and CD-46 Burkitt lymphoma cells, at par with those for colchicines. Cryptophycin 1 and 8, another anticancer compound was first isolated from *Nostoc sp.* by researchers at Merck.

The oral supplement of *Spirulina fusiformis* is known for regression of subjects with homogenous leukolakia (Mathew *et al.*, 1995). The extracts of *Spirulina* and *Dunaliella* inhibited the chemically induced carcinogenesis in model hamster buccal pouches (Schwartz *et al.*, 1988). Studies have also showed that sulphated polysachharide, calcium spirulans appears to inhibit tumor invasion of melanoma cells and basement membrane (Mishima *et al.*, 1998). *Aphanizomenon flosaquae* extract containing a high concentration of phycocyanin inhibited the *in vitro* growth of tumour cells, indicating the sensitivity of cell lines to the phycocyanin. A filamentous cyanobacterium *Phormidium tenue* contains several diacylglycerols that inhibit chemically induces tumors on mice (Tokuda *et al.*, 1996). Similarly, cryprophycin 1 isolated from *Nostoc* sp. (ATCC 53789) is the most potent suppressor of microtubule dynamics *i.e.* it blocks all cell cycle in G2/M phase. Curacin A is isolated from *Lyngbya majuscule.* This compound is found to be potent inhibitor of microtubule assembly.

Proteinase inhibitors

1. Aeruginosin 298 A isolated from *Microcystis aeruginosa* NIES 298- potent inhibitor of thrombin and trypsin (Matsuda *et al.*, 1996).
2. Microginins 299 A isolated from *M. aeruginosa* NIES 299 A- potent proteinase inhibitor (Ishida *et al.*, 1997b).

Antifungal activity

1. Flavanoid type compounds from *Oscillatoria* BDU 20801- anti-*Candida* activity.
2. Scytophytin from - *Scytonema pseudohofmanii.*
3. Laxaphycins from *Anabaena laxa* (Frankmolle *et al.*,1992).

Antimycobacterial activity of cyanobacteria

Rao *et al.* (2007) reported the Antimycobacterial Activity of different spp. of Cyanobacteria *viz., Hapalosiphon sp., Anabaena sp. Lyngbya sp., Westeillopsis prolifica, Spirulina sp. Anabaena variabilis, Anabaena cylindrica, Oscillatoria sp.* and *Scytonema sp.* against various species of Mycobacteria (*Mycobacterium tuberculosis ATCC 27294, M.tuberculosis MDR, M.avium, M. intracellulare and M. aurum).*

Anti-HIV Activity of Cyanobactria

Cyanobacterial species are also known to produce substances which have been proved to be anti-HIV, therefore can be exploited in therapy against AIDS. Gustafson *et al.* (1989) used a tetrazolum-based micro-culture to screen extracts of cultured marine cyanobacteria, *Lyngbya lagerheimii* and *Phormedium tenue*, for the inhibition of HIV-1. This led to the discovery of sulfonicacid containing glycolipids as a new class of HIV-1- inhibitory compounds. Other cyanobacteria, *Phormedium cebennse, Oscillatoria raciborskii, Scytonema burmanicum, Calothrix elenkinii* and *Anabaena variabilis*, gave extracts that inhibited HIV-1 and gave positive tests for the presence of sulfolipids.

Compound and extracts with anti-HIV activity are also active against other retroviruses such as Herpes simplex virus (HSV) and respiratory syncytial virus, but the amount of antiviral activity varies with the compound and the virus. Most of the research has focused on sulphated homopolysaccharides and heteropolysaccharides, sulfoglycolipids, carrageenans, fucoidan, sesquiterpene hydroquinones, and other classes of compounds with anti-HIV activity that have been isolated from algae have received less attention.

Hayashi *et al.* (1992) isolated calcium spirulan, a sulfated polysaccharide obtained from a marine blue green alga, *Spirulina platensis* which inhibited Herpes simplex virus. Subsequently, Ayehunie *et al.*, (1998) determined that an aqueous extract of *S. platensis*, at a concentration that were nontoxic to human cells, inhibited syncytium formation and HIV-1 replication in human T-cell lines, peripheral blood mononuclear cells and Langerhans cells. The antiviral effects of polysaccharide from marine algae towards mumps virus and influenza B virus were reported by Gerber *et al.*, (1958). Subsequently, polysaccharides fractions from extracts of red algae were found to inhibit herpes simplex virus (HSV). Similarly Boyd *et al.*(1997) isolated Cyanovirin-N from an aqueous cellular extract of cyanobacterium *Nostoc ellipsosporum* which has been proved to be antiviral. Lau et al., (1993) reported that the lipophilic and hydrophilic extracts of over 900 strains of cultured blue-green algae *in vitro* for their ability to inhibit the reverse transcriptases of avian myeloblastosis virus.

Table 2. Anti-HIV activity of compounds from Cyanobacteria

Cynobacterial strains	Compounds	References
Anabaena variabilis, Calothrix elenkinii, Lyngbya lagerheimii, Phormidium tenue, Phormedium cebennse, Oscillatoria raciborskii, Scytonema burmanicum.	Extracts containing sulfolipids (sulfoquinovosyl diacylglycerols)	Gustafson *et al.*, 1989
Spirulina platensis	activity in polysaccharide fraction	Ayehunie *et al.*, 1998

Contd...

Table Contd...

Spirulina platensis (Marine)	Calcium spirulan, Ca-SP (sulfated polysaccharide fraction) Dextran sulfate	Hayashi *et al.*, 1996
Microcystis aeruginosa	Aqueous extract showed antiviral activity against influenza A virus	Nowotny *et al.*, 1997
Nostoc ellipsosporum	Cyanovirin- N (11-k Da antiviral protein	Gustafson *et al.*, 1996
Nostoc ellipsosporum	Cyanovirin- N (11-k Da antiviral protein	Boyd *et al.*, 1996
900 strains	Lipophilic and Hydrophilic extracts	Lau *et al.*, 1993
Scytonema spp.	Sulfoglycolipids	Rashef *et al.*, 1997
Oscillatoria spp.	Acylated diglycolipids	Rashef *et al.*, 1997
Oscillatoria raoi (TAU IL-76-12), *Syctonema spp.* (TAU SL-30-1-4), *Oscillatoria trichoides* (TAU IL 104-3-2), *Phormidium tenue* (TAU IL-144-1), *Oscillatoria limnetica Lammermann* (TAU NG-4-1-2)	Sulfolipids, Sulfoglycolipids, hydrolysis products, synthetic derivative	Loya *et al.*, 1998

Anti-HIV activity of algae other than cyanobacteria

Compounds of the extracts from algae have *in vitro* and *in vivo* activity against a wide range of retroviruses including Herpese viruses, Togaviruses, Paramyxoviruses, Rhabdoviruses and Human immune deficiency viruses (HIV). Componds isolated from algae that have been tested against HIV include steroids and sulfoglycolipids, but most of the research had used natural and synthetic sulfated polysaccharide. Most of the research on the anti-HIV activity of marine algae has used red and brown microalgae. The initial studies using these algae isolated sulphated polysaccharides with antiviral activity and later investigation continued interest in the class of compounds (Table 3).

Table 3. Anti-HIV activity of extracts and compounds isolated from algae other than cyanobacteria.

Species	Compound Isolated	References
Fucus vesiculosus (brown seaweed)	Polysachharide, Fucoidan, polyphenols	Beress *et al.*, 1993
Fucus vesiculosus (brown seaweed)	Fucoidan and water soluble, noncarbohydrate component of fucoidan	Moen and Clark, 1993
Sargassum horneri (brown seaweed)	Sulfated Polysaccharide with fucose as the main component sugar	Hoshino *et al.*, 1998
Cochlodinium polykrikoides	Sulfated Polysaccharide	Hasui *et al.*, 1995
Nothogenia fastigiata (red seweed)	Sulphated xylomannan	Damonte *et al.*, 1994
Nothogenia fastigiata, Asardheiella tenera, Eucheuma cottonii, Gigartina aciculaire, Gigartina pistillata (Marine algae)	Xylomannan, Galactan sulfate, lamda Carrageenan	Witvrouw and De Clercq , 1997
Schizymenia pacifica (Red seweed)	Extracts, polysachharide MW, 102 kDa	Nakashima *et al.*, 1987
Gigartina tenella (red seaweed)	Sulfoquinovosyl-diacylglycerol	Ohta *et al.*, 1998
Peyssonelia sp. (red seaweed)	Peyssonol A, Peyssonol B (sesquiterpene hydroquinones)	Loya *et al.*, 1995
Purchased compound rather than isolated	Carrageenan, Fucoidan	Bagasra *et al.*, 1989

Keeping in view the above consideration and significance of secondary metabolites from cyanobacteria, the present work was carried out to evaluate the antibacterial activity of selected species of cyanobacteria against some selected pathogenic microoraganisms *viz. Staphylococcus aureus, Escherichia coli, Pseudomonas aeruginosa, Klebsiella pneumoniae* and *Salmonella typhi.*

□□□

Chapter 2

REVIEW OF LITERATURE

Cyanobacteria have been identified as one of the most promising groups of organisms from which novel and biochemically active natural products are isolated. Cyanobacteria such as *Microcystis, Anabaena, Nostoc* and *Oscillatoria* produce a great variety of secondary metabolites. The only group comparable to cyanobacteria is actinomycetes, which has yielded a tremendous number of secondary metabolites. The rate of discovery from traditional microbial drug producer like actinomycetes, which is in the focus of pharmaceutical research for decades, is decreasing and it is the time to turn to cyanobacteria and exploit their potential. Recently, there has been an increasing interest in cyanobacteria as a potential source of new drugs (Glombitza and Koch, 1989; Schwartz *et al.*, 1990). This is of paramount importance to fight increasingly resistant pathogens and newly emergent diseases (Hayashi *et al.*, 1994).

Cyanobacteria are photosynthetic microorganisms that have flourished on earth for more than 3 billion years with some genera showing only minor morphological changes since that time (Hagmann and Juttner, 1996). The medicinal and nutrient qualities of cyanobacteria were first appreciated as early as 1500 BC, when *Nostoc* species were used to treat gout, fistula and several forms of cancer (Pietra, 1990; Bladon, 2002; Liu and Chen, 2003).

Secondary metabolites refer to those compounds that are not used by the organisms for its primary metabolisms. Secondary metabolites influence other organisms in the vicinity and are thought to be of phylogenetic importance. Secondary metabolites include several types of compounds that may act as harmones, antibiotics, allelochemicals, toxins, and biotoxins that are found in surface supplies of freshwater (Carmichael, 1992; Patterson *et al.*, 1994; Moore, 1996; Smith and Doan, 1999). The ability of such compounds to kill bacteria and fungi has been well documented (Bonjouklian *et al.*, 1988). The properties of secondary metabolites in nature are not completely understood (Metting and Pyne 1986; Inderjit and Dakshini, 1994).

In a review by Metting (1996), an estimate was made for the number of species of cyanobateria in existence and many of these species have strains that produce large variety of different metabolites. Several compounds like tannins, bromophenols, terpenols, and polysaccharides are secreted by cyanobacteria proven to have antibiotic properties (Fatma and Venkataramall, 1994).

Cyanobacteria provide a potential source of biologically active secondary metabolites including antibacterial (Ishida *et al.*, 1997; Mundt *et al.*, 2001; Rezanka and Dembitsky 2006), antifungal (MacMillan *et al.*, 2002), cytotoxic (Luesch *et al.*, 2000), algicidal (Papke *et al.*, 1997), immunosuppressive (Koehn *et al.*, 1992), and antiviral activities (Hayashi and Hayashi, 1996). Schaeffer and Krylov (2000) reported the anti-HIV activity of extracts and compounds from algae and cyanobacteria.

The first partly identified antimicrobial compound isolated from algae were obtained from unicellular green algae particularly, *Chlorella* which contained a substance termed as '*chlorellin*' that exhibited inhibitory activity against both Gram-positive and Gram-negative bacteria, including *Staphylococcus aureus, Streptococcus pyogenes, Bacillus subtilis* and *Pseudomonas aeruginosa* (Pratt *et al.*, 1944). Chlorellin is composed of peroxides of unsaturated fatty acids (Spoehr *et al.*, 1949).

Since 1981, a team of workers at the University of Hawaii, along with colleagues at other institutions in certain instances, have sought to discover new compounds of pharmaceutical interest from cyanobacteria. To this end, they have evaluated for biological activity, hydrophilic and lipophilic extracts from more than 1500 strains that represented approximately 400 species of cyanobacteria from terrestrial, freshwater, and marine habitats (Patterson *et al.,* 1994). Antifungal and antiviral activity was observed in a large number of these extracts.

A number of cyanobacteria and eukaryotic algae are known to produce antibiotics that significantly inhibit various bacterial and fungal strains in laboratory tests (Kulik, 1995). Filamentous cyanobacterium, *Scytonema sp.* TISTR 8208 produces a cyclic peptide antibiotic (Chestsumon, 1995).

Pratt *et al.* (1951) reported that extracts from several species of seaweed inhibited the growth of *S. aureus, E. coli, P. aeruginosa.* Burkholder *et al.* (1960) reported that 66 out of 131 identified species showed some degree of inhibitory activity towards one or more of the microbial strains mainly against *S. aureus, E. coli* and *Mycobacterium smegmatis* and to *Candida albicans.* In a similar study, Welch (1962) tested homogenized preparation of 32 marine algae for activity against six fungi. Four of the preparation severly inhibited the growth of the *Rhizopus oryzae, Mucor racemosus, Aspergillus niger, Penicillium sp., Candida albicans* and *Cryptococcus neoforman.*

Sieburth (1965) reported that senescencing bloom of *Skeletonema costatum* showed anticoliform activity and also found that concentrated acetone extracts of this diatom inhibited marine isolates of *Vibro* and *Pseudomonas* as well as being active against *E. coli* and *Sarcina lutea,* whereas other Gram-positive and Gram-negative terrestrial bacteria were virtually unaffected.

Ramamurthy (1970) obtained indirect evidence that *Trichodesmium erythraeum;* a marine cyanophyte produced an antibiotic that was effective against Gram-positive and Gram-negative bacteria.

Sims *et al.* (1975) reported the antimicrobial activity of five compounds extracted from marine algae, which were tested against *S. aureus, S. chleraesuis, Mycobacterium smegmatis, Candida albicans* and *E. coli.* Three of the compounds namely cycloeudesmol, laurinterol and debromolaurinterol exhibited activity at concentration approaching that of streptomycin. The most active compound of five tested was laurinterol, which gave complete inhibition of *S. aureus* and *M. smegmatis.* Only one compound, cycloeudesmol significantly inhibited a Gram-negative bacterium, however, debromolaurinterol showed greater activity against *C. albicans* as compared to laurinterol.

Table 4. Secondary Metabolites recently described from Cyanophytes.

Cyanobacteria	Active compound	Effect	References
Anabaena circinalis	Circinamide	Papain inhibitor	Shin *et al.*, 1997
Anabaena cylindrica	Dehydroradiosumin	Trypsin inhibitor	Kodani *et al.*, 1998
Dichaothrix baueriana	β-carbolines	Antiviral	Larsen *et al.*, 1994
Fischerella ambigua	Polyhalogenated Aromatic compound	Cytotoxic	Falch *et al.*, 1992
Fischerella muscicola	Fischerellin A	Fungicide	Hagmann & juttner, 1996
Lyngbya lagerheimii	Sulfolipidis	Antiviral"(HIV-1)	Gustafson *et al.*, 1989
Lyngbya majuscula *Schizothrix calcicola*	Lyngbyastatin 1	Cytotoxic	Harrigan *et al.*, 1998b
Lyngbya majuscula	Curacin A	Anti-tumor	Lai *et al.*, 1996
Lyngbya majuscula	Lyngbyabrllin A (1)	Cytotoxic	Luesch *et al.*, 2000
Microcystis aerugionosa	Lipid	Algaecide	Ikawa *et al.*, 1996.
Microcystis aerugionosa	Micropeptin 478-A Micropeptin 478-B	Plasmin inhibitor	Ishida *et al.*, 1997a

Contd...

Table Contd...

Microcystis aerugionosa	Microginin 299-A Microginin 299-B	Leucin aminopeptidase inhibitor	Ishida *et al.*, 1997b
Microcystis aerugionosa	Kawaguchipeptin	Bactericide	Ishida *et al.*, 1997c
Microcystis aerugionosa	Aqueous extract	Antiviral(influenza A)	Nowotny *et al.*, 1997
Microcystis viridis	Aeruginosin 102A Aeruginosin 102A	Thrombin inhibitor	Matsuda *et al.*, 1996
Microcystis viridis	Micropeptin 103	Chymotrypsin inhibitor	Murakami *et al.*, 1997 a
Nodularia spumigena	Nodulapeptin A Nodulapeptin B	Protracted toxic	Fujii *et al.*, 1997
Nostoc sp.	Cryptophycins	Anti-tumor	Trimurtulu *et al.*, 1994
Nostoc sp.	Bacteriocin	Antibacterial	Flores & Wolk, 1986
Nostoc commune	Nostodione A	Antimitotic	Kobayashi *et al.*, 1994
Nostoc ellipsosporum	Cyanovirin N	Antiviral(HIV)	Boyd *et al.*, 1997
Nostoc minutum	Microviridin	Clastase inhibitor	Murakami *et al.*, 1997a
Nostoc minutum	Nostopeptin A Nostopeptin B	Clastase inhibitor	Okino *et al.*, 1997

Contd...

Table Contd...

Nostoc muscurum	Phenolic compounds	Fungicide	De Cano *et al.*, 1990
Nostoc sphaericum	Indolocarbozole	Antiviral (herpes simplex, virus type 2)	Falch *et al.*, 1995
Nostoc spongiaeforme var. *tenue*	Tenuecyclamides A-D	Growth inhibitor	Banker & Carmeli, 1998
Oscillatoria agardhii	Oscillapeptin G	Tryrosinase inhibitor	Sano & Kaya, 1996
Oscillatoria agardhii	Anabaenopeptin B	Protease inhibitor	Murakami *et al.*, 1997c
Oscillatoria limnetica	Glycolipids	Antiviral (HIV-1)	Reshef *et al.*, 1997
Oscillatoria raoi	Acetylated sulfoglycolipids	Antiviral	Reshef *et al.*, 1997
Oscillatoria trichoides	Glycolipids	Antiviral (HIV-1)	Loya *et al.*, 1998
Phormidium tenue	Sulfolipids	Antiviral (HIV-1)	Gustafson *et al.*, 1989
Phormidium tenue	Sulfolipids	Antiviral (HIV-1)	Falch *et al.*, 1995
Scytonema ocellatum	Phytoalexin	Fungicide	Patterson & Bolis, 1997
Spirulina platensis	Calcium spirulan	Antiviral	Hayashi *et al.*, 1996
Stigonema dendroideum	Dendroamids	Reversing multidrug resistance	Ogino *et al.*, 1996
Symploca hydnoides	Symplostatin 1	Cytostatic	Harrigan *et al.*, 1998a
Tolypothrix conglutinate	Tolytoxin	Cytotoxin and fungicide	Moore *et al.*, 1990

Contd...

Khaleafa *et al.* (1975) reported that the ethanol extracts from three out of 18 marine algae had a pronounced inhibitory effect on unidentified leather moulds. Accorinti (1983) listed 35 marine and 10 freshwater algae that produced substance that inhibited the growth of *Candida albicans* and several other fungi. Matusiak and Krzywicka (1975) reported that an extract from *Chlorella vulgaris,* a terrestrial chlorophyte, inhibited the growth of *Aspergillus oryzae, Saccharomyces cerevisiae* and the human pathogen *Trichosporon cutaneum.*

Kobbia and Zaki (1976) found that the filtrate from *Brachytrichia balani* inhibited the growth of *Staphylococcus aureus* and *Salmonella sp.* and that from *Mastigocoleus testarum* greatly inhibited *Escherichia coli, Salmonella sp.* and *Mycobacterium phlei.*

The terrestrial chlorophyte *Chlorococcum humicolum* produced a broad spectrum antibiotic which was quite inhibitory to *Cryptococcus neoformans, Bacillus subtilis, Escherichia coli* and *Staphylococcus aureus* and to saprophytic bacteria *Bacillus subtilis* and *Sarcina lutea* according to Pande and Gupta (1977).

Debro and Ward (1979) studied the antibacterial activity of ethanol extracts of 24 species of green algae against Gram-positive and Gram-negative bacteria. Fourteen algae exhibited significant inhibitory activity against *Bacillus subtilis* and *Staphylococcus aureus.* No activity was found against *Escherichia coli.*

Lyngbya majuscula was reported to produce an antibiotic, malyngolide that was effective against *Mycobacterium smegmatis* and *Streptococcus pyogenes* but less effective against *Staphylococus aureus* and *Bacillus subtilis* (Cardellina *et al.,* 1979).

Moon and Martin (1981) reported that a crude extract of *Gomphosphaeria aponina,* a marine cyanophyte inhibited the yeast *Saccharomyces cerevisiae.* Wen and Weici (1984) reported the antimicrobial activity of 60 species of marine algae from the Qingdao coast, China against Gram-positive (*Bacillus subtilis*), Gram-negative (*Escherichia coli*), Marine bacterium (*Vibrio sp.*) and one yeast (*Saccharomyces cerevisiae*).

Espeche *et al.* (1984) reported the antibacterial activity of crude dialyzed extract of argentine marine algae against *Bacillus subtilis, Sarcina lutea, Staphylococcus aureus* and *Escherichia coli.*

Rosell and Srivastava (1987) reported the antibacterial activity of nine brown algae against nine bacterial species by disc diffusion method. They have found that all the extracts showed antibacterial activity against several Gram-negative (*Branhamella catarrhalis, Escherichia coli, Proteus vulgaris, Pseudomonas fluorescens* and *Serratia marcescens)* and Gram-positive bacteria (*Bacillus cereus, Bacillus subtilis, Micrococcus luteus* and *Staphylococcus aureus*).

Lipid soluble extracts from 102 species of Caribbean marine algae were tested for antibiotic activity by Ballantine *et al.* (1987). Inhibition zones were ranged in size from 5.0 to 6.0 mm. Among all the extracts examined, 65% of them demonstrated at least minimal activity against one or more of four bacterial and one fungal species. 61% of the extracts showed activity only against *Bacillus subtilis* and *S. aureus.* 19% extracts were found active only against *Candida albicans* and the 50% were active against *P. aeruginosa* and *E. coli.*

Ballantine *et al.* (1987) mentioned that a number of species in their study showed temporal variability in their microbial activity. This included inactivity at one sampling time, activity against one set of bioassay microorganisms at a second sampling and activity against a different set when sampled for the third time. They cited following factor that influence the antimicrobial activity of algal extracts: reproductive state, sampling location and seasonality.

Cultures and organic solvent extracts of 100 cyanobacteria and 300 eukaryotic freshwater algae were screened for their antibacterial activity against *Bacillus subtilis, Escherichia coli* and *Staphylococcus aureus* by Cannell *et al.* (1988). Activity was detected in samples from 36 eukaryotic and five cyanobacterial cultures. Only methanolic extract was found active against *B. subtilis.*

More than 700 clonal isolates of cyanobacteria from various terrestrial, freshwater, and marine habitats were mass-cultured by Stewart *et al.* (1988). Extracts from about 9% of these cultures exhibited antifungal activity against some of the test fungi including *Aspergillus oryzae, Candida albicans, Penicillium notatum, Saccharomyces cerevisiae,* and *Trichophyton mentagrophytes.* Culture filtrate and organic solvent extracts of 132 marine and 400 freshwater algal cultures were evaluated by Kellam *et al.* (1988) for inhibitory properties against six fungi. Their results indicated that marine algae are a more promising source of antifungal agents than fresh water algae.

Kellam *et al.* (1988) has reported that extract obtained from *Anabaena variabilis*, a fresh water cyanobacterium severely inhibited the growth of the cellulolytic fungus *Chaetomium globosum.* Gerwick *et al.* (1989) isolated a complex mixture of toxic and antimicrobial peptides from the marine cyanophyte *Hormothamnion enteromorphoides.* The major peptide named hormothamnin A was found to be active against *Bacillus subtilis, Escherichia coli, Pseudomonas aeruginosa, Salmonella typhimurium* and *Staphylococcus aureus.* In addition, this compound was active against *Candida albicans* and *Trichophyton mentagrophytes.*

Phenolic compounds were also determined in methanolic extract from the algal mass of a *Nostoc muscorum* culture. Bioassay with two human pathogen, *Candida albicans* and *Staphylococus aureus* indicated that algal phenolic compounds evoked significant growth inhibition for both species (89.1% and 88.25%, respectively). It is suggested that this strong inhibitory effect is of potential medicinal value (De Cano *et al.*, 1990).

Patterson *et al.* (1991) obtained extracts from approximately 1,000 strains of cyanobacteria from various habitats and screened them for possible antineoplastic activity. The following cytotoxic antibiotics, produced by 22 strains from seven genera, unique to cyanobacteria, were discovered: acutiphycins, indolecarbazoles, mirabilene isonitriles, paracyclophanes, scytophycins, tantazoles, tolytoxin, toyocamycin, and tubercidin.

Bonjouklian *et al.* (1991) found that a lipophilic extract from *Tolypothrix tjipanasensis* (De Wild) had moderate fungicidal activity against *Candida albicans, Trichophyton mentagrophytes,* and *Aspergillus flavus.* The structures of 15 fungicidal alkaloids from the extract were determined, and were given the name tijpanzoles. Extracts of four marine algae were reported by Tariq (1991) and were demonstrated to inhibit the lateral growth of colonies of the human pathogenenic fungi *Microsporum canis* and *Trichophyton verrucosum.*

Ballesteros *et al.* (1992) tested extracts from 71 species of marine macrophytes and found that 70% of them demonstrated antifungal activity, 21% antiviral activity and 6% antibacterial activity. The highest level of antibiotic activity came from extracts from members of chlorophyta. Frankmolle *et al.* (1992) reported that crude ethanolic extracts from *Anabaena laxa* inhibited the growth of the following fungi:

Aspergillus oryzae, Candida albicans, Penicillium notatum, Saccharomyces cerevisiae and *Trichophyton mentagrophytes*. The fungicidal compounds were isolated and purified and given the name laxaphycins A, B, C, D, and E and their structures were also determined.

Moon *et al.* (1992) isolated and determined the structure and fungicidal activity of a compound named calophycin, produced by *Calothrix fusca*, this compound was effective against *Aspergillus oryzae, Candida albicans, Penicillium notatum, Saccharomyces cerevisiae,* and *Trichophyton mentagrophytes*.

Smitka *et al.* (1992) isolated six hapalindole–type alkaloids with fungicidal properties from *Fischerella ambigua, Hapalosiphon hibeornicus* and *Westiellopsis prolofica*. These compounds were identified, as ambiguine isonitriles A-F and their structure were determined. They inhibited the growth of the five test fungi (Moon *et al.*, 1992).

Chen *et al.* (1993) determined the structure of two antibacterial compounds, gloiosiphones A and B, isolated from the marine algae *Gloiosiphonia verticillaris*. Crude extracts of these compounds were active against *Bacillus subtilis, Escherichia coli, Salmonella typhimurium* and *Staphylococcus aureus*. Bernart *et al.* (1993) isolated a novel fatty acid derivative from the marine algae *Acrosiphonia coalita* that inhibited the growth of *Bacillus subtilis, Candida albicans* and *Staphylococcus aureus*.

Dunaliella primolecta C-525, exhibiting the highest antibiotic activity and was selected out of 84 marine algae tested. 35 strains of cynophyceae, 28 strains of Chlorophyceae, 11 strains of Bacillariophyceae, four strains of Dinophyceae, three strains of Rhaphidophyceae, two strains of Rhodophyceae and one strain of Chlorophyceae in the crude extract were tested against a Gram-negative bacterium, *Escherichia coli* and a Gram-positive bacterium *Staphylococcus aureus*. Although all crude extracts of the 84 strains exhibited no activity towards *E. coli*, the crude extract of seven strains exhibited strong antibiotic activity against *S. aureus, Bacillus subtilis* and *Enterobacter aerogenes* (Chang *et al.*, 1993).

Ohta *et al.* (1993) reported the antibiotic activity of crude extracts of 68 algal isolates against a Gram-positive bacterium, *Staphylococcus aureus* and a Gram-negative bacterium, *Escherichia coli*. Crude extracts of five algal isolates inhibit the growth of *S. aureus*. No activity was observed against *E. coli*.

Fish and Codd (1994) reported that a thermotolerant species of *Phormedium* produced extracellular antimicrobial material during batch culture; it inhibited the growth of a wide range of Gram-positive and Gram-negative heterotophic bacteria, *Candida albicans* and *Cladosporium resinae*. Although this material was inactive when screened against a number of other cyanobacteria.

Gerwick *et al.* (1994) over a three year period evaluated more than 500 extracts from marine micro- and macroalgae for anticancer activity. Most of the active compounds (67%) were obtained from the later. Chloroform extracts from the marine algae *Gracilaria corticata* and *Padina tetrastromatica* inhibited the growth of *Escherichia coli, Proteus vulgaris, Pseudomonas aeruginosa, Salmonella paratyphi, S. typhi* and *Staphylococcus aureus*.

Falch *et al.* (1995) investigated a total of 80 lipophilic and hydrophilic extracts obtained from 20 samples of cultured freshwater and terrestrial cyanobacteria. Out of 54 extracts tested for antimicrobial activity, 78% showed antibacterial and 45% showed antifungal activity. They used successive extractions with solvents of increasing polarity *viz*. petroleum ether, dichloromethane, methanol and 80% aqueous methanol and water. These extracts showed different antibacterial effect in bioautographic assays with *Bacillus subtilis* and *Micrococcus luteus*. Bioassay-guided fractionation of the lipophilic extracts of *Fischerella ambigua* led to the isolation of three compounds: ambigols A(1) and B(2) and tjipanazole D(3). Compounds 1 and 2 exhibited antibacterial, antifungal, cytotoxic, molluscicidal, anti-inflammatory and antiviral activity. Tjipanazole D showed moderate antibacterial properties.

Antibacterial substances were isolated and purified from the red algae (*Laurencia okamurae* collected from Kashiwazima, Japan, in June 1995) by successive silica gel chromatography and silica gel TLC. MIC of the antibacterial substance LO-B was 12.5 g/disc to all Gram-positive bacteria tested. MIC of LO-C was 1.56 gm/disc to bacteria *Streptomyces acidiscabies* and to methicillin resistant *Staphylococcus aureus* (Horikawa *et al.*, 1996).

Mtolera and Semesi (1996) reported the antimicrobial activity of six green algae (*Valonia aegrophila, Ulva pertusa, Halimeda opuntia, Halimeda, Caulerpa racemosa* and *Caulerpa mexicana*) from Tanzania against three bacterial species *viz., Staphylococcus aureus, Bacillus subtilis, Escherichia coli*, and one yeast, *Candida albicans* using the disc diffusion assay. They have concluded that all the algal extracts except *Caulerpa mexicana* were found to be active.

Centeno *et al.* (1996) evaluated the antibacterial activity of lipid soluble extracts from three species of Caribbean marine algae, *Spyridia filamentosa, S. hypnoides* and *Wrangelia bicuspidate* monthly for one year against *Staphylococcus aureus, Bacillus subtilis, Escherichia coli* and *Pseudomonas aeruginosa*. Activities were tested for whole plant extract and for TLC-separated zones. Whole plant extracts demonstrated monthly variability in activity with respect to both the habitat and life history phase in addition to periods of similarity. No consistency was seen regarding activity against different microorganisms.

Stensvik *et al.* (1998) reported the antibacterial activity of aqueous and methanol extracts from five selected cyanobacteria in six different bacterial bioassays. All five cyanobacteria revealed antibacterial properties. They had observed that methanolic extract made from *Trychonema bourrellyi, Aphanizomenon flos-aquae* and *Cylindrospermopsis raciborskii* showed the most pronounced inhibitory effects whereas aqueous extracts made from *Microcystis aeruginosa* and *T. bourrellyi* possessed evident antibacterial properties.

Jaki *et al.* (1999) screened a total of 86 lipophilic and hydrophilic extracts obtained from 43 samples of cultured and field-collected fresh water and terrestrial cyanobacteria for their biological activities. 16.3% showed activity against Gram-positive bacteria and 5.8% against Gram-negative while 10.5% possessed antifungal activity.

Kreitlow *et al.* (1999) isolated hydrophilic and lipophylic extracts of twelve cyanobacterial strains from fresh and brackish water blooms, collected during the summer from the Baltic Sea, were investigated for their antibacterial activity against seven microorganisms. They have reported that extracts were active against Gram-positive bacteria (*S. aureus, Micrococcus flavus* and *B. subtilis*) whereas, no inhibitory effect was observed against the Gram-Negative bacteria (*E. coli, Proteus mirabilis, Serratia marcescens* and the yeast, *Candida maltosa).*

Jaki *et al.* (2000) isolated two new compounds, a diterpenoid and an anthraquinone, as well as an indane derivative from the cells of the cultured cyanobacterium *Nostoc commune* (EAWAG 122b), all the compounds were found active against *S. epidermidis* and *B. cereus*. They also have determined the structure by spectroscopic methods mainly NMR, infrared spectroscopy and mass spectroscopy. They have further opined the MICs of diterpenoid, anthraquinone and indane derivative were 4, 32, and 128 (in ppm) against *S. epidermidis* and 128, 32, 64 (in ppm) against *B. cereus,* respectively.

Heyduck and Fischer (2000) reported that cyanobacteria produce extracellular substance with different biological activity and behaviour. Excreted substances from *Anabaena sp.* strain Hi 26, *Oscillatoria subtilissima* strain Bo 62, *Rivularia sp.* strain Bo 85 and *Oscillatoria limnetica* strain Flo 1 affect in varying degrees the growth of Gram-positive and Gram-negative heterotrophic bacteria, as well as of the yeast *Sacharomyces cerevisiae.*

Piccardi *et al.* (2000) reported the antimicrobial properties of hydrophilic and lipophilic extracts of fifty cyanobacterial strains from different habitats (symbiosis, soil, fresh and marine waters) belonging to the genus *Nostoc.* They have tested both the crude extracts for antifungal (against *Penicillium expansum* and *Rhizoctonia solani)* and antibacterial activity (against *Agrobacterium vitis, Escherichia coli* and *Staphylococcus epidermidis).* Twenty four strains showed activity against at least one of the target organisms.

Doan *et al.* (2000) isolated 12-epi-hapaliadoli E-isonitrile from *Fischerila sp.* and the indolophenan calothrixin A from *Calothrix sp.* that inhibited RNA synthesis in *Bacillus subtilis.*

Filho *et al.* (2002) reported the antibacterial activity of Hexane, chloroform and ethanol extracts of six marine microalgae (rhodophyta and chlorophyta) from North Ceara coast (Northern Brazil) against Gram-positive and Gram-negative Bacteria. They found that the best result were shown by the hexane extracts of *Amansia multifida* against *Enterobacter aerogenes, Klebsiella pneumoniae, Pseudomonas aeruginosa, Salmonella typhi, S. choleraesuis, Serratia marcesscens, Vibrio cholerae, Bacillus subtilis* and *Staphylococcus aureus.*

The antibacterial n-hexane extract of *Limnothrix redkei* HUB 051 was fractionated using different separation techniques. Two active fractions were received and one was identified by NMR and MS as α-*linolenic acid.* The antibacterial activity of this polyunsaturated fatty acid is well known. In 1981, McDonald and coworkers reported about antibacterial effect of α-*linolenic acid* against methicillin resistant *Staphylococcus strains.* A Japanese group isolated α-*linolenic acid* as the antibiotic principle from the biomass of green alga *Chlorococcum sp.* HS-101 (Ohta *et al.,* 1995). The MICs for α-*linolenic acid* range from 75 mg/ml against *Staphylococcus aureus* SBUG 11 to 25 mg/ml against *Micrococcus flavus* SBUG 16, but ampicillin was more effective against these bacteria. The MICs were 0.05 mg/ml and 0.25 mg/ml, respectively.

Pure cultures of *Microcystis aeruginosa, Platymonas viridis and Nephrochloris salina* have been grown on the media with different nitrogen and phosphorus content. Their supernatants and pellets, as well as lipid complex, terpene fraction and some components from *M. aeruginosa* have selective antibacterial characteristics. The increase of nitrogen content in the medium correlated with intensification (*M. aeruginoas, N. salina*) or conservation (*P. viridis*) of bactericidal activity. The pellet fraction was more active than supernatant one. The specific cyanobacterial and microalgal inhibitory effect is supposed with respect to the organisms of different evolutionary level (Gol'din, 2003).

Ghasemi *et al.* (2004) isolated *Fischerella anbigua* during a screening program from soil sample collected from paddy fields in the north of Iran. It was shown that it excreted broad-spectrum antibacterial and antifungal substances in culture medium of the *F. ambigua* led to the isolation of novel antimicrobial substances, which was named parsiguine that exhibited antifungal and antibacterial activity. They also have determined the structure by spectroscopic methods, including NMR, MS, IR and UV and the MIC values were in the range of 40 µg/ml against *S. epidermidis* and 20 µg/ml against *C. krusei*.

Ozdemir *et al.* (2004) reported the antimicrobial activity of methanol, dichloromethane, petroleum ether, ethyl acetate extracts and volatile components of *Spirulina platensis* against four Gram-positive, six Gram-negative and *Candida albicans*. They have observed that methanol extract showed more potent activity than dichloromethane, petroleum ether, ethyl acetate and volatile components. GC-MS analysis of the volatile components of *Spirulina platensis* resulted in the identification of 15 compounds which constituted 96.45% of the total compounds. The volatile components of *Spirulina platensis* consisted of heptadecane (39.7%) and tetradecane (36.61) as major components.

Pelegrin and Morales (2004) reported the antibacterial activity of ethanolic and lipid-soluble extracts from 21 marine algal species (10 chlorophyta, 2 phaeophyta and 9 rhodophyta) from the coast of Yucatan, Mexico against pathogenic Gram-positive, Gram-negative and one fungus. Activity observed only against Gram-positive bacteria (*Bacillus subtilis, Streptococcus faecalis* and *Micrococcus luteus)*

Oranday *et al.* (2004) reported the antimicrobial properties of marine algae (*Gracilaria tikvhiae, Ulva lactuca, Ulva fasciata* and *Sargassum fluitans)* against *Escherichia coli, Staphylococcus aureus, Pseudomonas aeruginosa, S. enteriditis, S. epidermidis, Streptococcus faecalis* and *Candida albicans* by disc diffusion method. They have found that non-polar extracts of *Sargassum fluitans* and polar extract of *Gracilaria tikvhiae* inhibited the growth of more than four microorganisms. Extract were separated using chromatography column and fraction were tested against *S. aureus* and *C. albicans*. They have also reported the minimum inhibitory concentration of eight fraction of petroleum ether of *S. fluitans* against *C. albicans* (0.16 μg/ml).

Soltani *et al.* (2005) reported the antibacterial properties of aqueous, petroleum ether and methanolic extracts from 76 microalgae isolated from paddy field of seven provinces in Iran against four bacteria and two fungi. Out of total microalgae, 17 cyanobacteria exhibited antimicrobial effect.

Choudhury *et al.* (2005) reported the antibacterial activity of three marine algae *viz., Gracilaria corticata, Ulva fasciata* and *Enteromorpha compressa* against six virulent strain of bacteria pathogenic to fish *viz. Edwardsiella tarda, Vibrio alginolyticus, Pseudomonas fluorescens, Pseudomonas aeruginosa* and *Aeromonas hydrophila* (2 strains).

Safonova & Reisser (2005) reported that supernatant from the culture of the taxa of chlorophycean, trebouxiophycean & xanthophycean soil microalgae and cyanobacteria (*Chlorococcus turgidus, Nostoc sp.* and *Synechocystis sp.)* have been tested for the release of substance that inhibit the growth of *E. coli* and *Micrococcus luteus.*

Asthana *et al.* (2006) isolated *Fischerella sp.* from Neem (*Ajadirachta indica*) tree bark which was found active against Gram-positive and Gram-negative bacteria with the highest zone of inhibition (40mm) against *S. aureus*. Based on MS, UV, IR, & NMR analyses, the activity was supposed due to hapalindole T having the empirical formula $C_{22}H_{23}N_2ClSO$ and molecular weight of 386 with the melting point range 179-182^{0}C. They also have determined the MICs of the active principal against *S. aureus* (0.25 μg/ml), *P. aeruginosa* (2.0 μg/ml), *E. coli* (4.0 μg/ml) and *S. typhi* (4.0 μg/ml).

Table 5: *In vitro* antimicrobial activity of reported cyanobacteria

Cyanobacteria	Microorganisms	Active Compound	References
Anabaena lemmermannii	*Bacillus subtilis, Bacillus cereus.*	NI	Stensvik *et al.*, 1998
Anabaena planctonica 115b	*Bacillus subtilis, Micrococcus luteus, Escherichia coli, Penicillium oxalicum*	NI	Falch *et al.*, 1995
Anabaena sp.	*Micrococcus flavus, Bacillus megaterium, Bacillus thuringiensis*	NI	Kreitlow *et al.*, 1999 & Katircioglu *et al.*, 2006
A. solitaria	*Bacillus subtilis, Micrococcus flavus*	NI	Kreitlow *et al.*, 1999
Anabaena variabilis 1	*Bacillus subtilis, Chaetomium globosuim*	NI	Mundt *et al.*, 2001 Kellam *et al.*, 1988
Anabaena laxa	*Aspergillus oryzae, Candida albicans, Penicillium notatum, Saccharomyces cerevisiae, Trichophyton mentagrophytes*	Laxaphycins	Frankmolle *et al.*, 1992 a, b
Aphanizomenon flos-aquae	*Bacillus subtilis, Bacillus cereus, Escherichia coli*	NI	Stensvik *et al.*, 1998
Aphanizomenon flos-aquae 204	*Bacillus subtilis, Micrococcus luteus, Escherichia coli*	NI	Falch *et al.*, 1995
Brachytrichia balani	*Salmonella sp., Staphylococcus aureus*	NI	Kobbia and Zaki, 1976
Calothrix fusca	*Aspergillus oryzae, Candida albicans, Penicillium notatum, Saccharomyces cerevisiae, Trichophyton mentagrophytes*	Calophycin	Moon *et al.*, 1992
Cylindrospermopsis raciborskii	*Bacillus subtilis, Bacillus cereus.*	NI	Stensvik *et al.*, 1998

Contd...

Table Contd...

Chrococcus sp.	*Bacillus subtilis, Pseudomonas aeruginosa, Candida tropicalis*	NI	Katircioglu *et al.*, 2006
Chrococcus sp. FS1	*Bacillus subtilis*	NI	
Cylindrospermum majus 95	*Bacillus subtilis*	NI	Mundt *et al.*, 2001
Calothrix gracilis 96	*Bacillus subtilis*	NI	Mundt *et al.*, 2001
Fischerella ambigua	*Aspergillus oryzae, Candida albicans, Penicillium notatum, Saccharomyces cerevisiae, Trichophyton mentagrophytes*	Ambiguines	Smitka *et al.*, 1992
Fischerella ambigua 108a	*Bacillus subtilis, Micrococcus luteus, Escherichia coli*	Ambigol A & B	Falch *et al.*, 1993 & 1995
Fischerella ambigua	*Staphylococcus epidermidis, Candida krusei*	Parsiguine	Ghasemi *et al.*, 2004
Fischerella sp.	*Mycobacterium tuberculosis, Enterobacter aerogenes, Staphylococcus aureus, Pseudomonas aeruginosa, Salmonella typhi, E. coli.*	Hapalindole	Asthana *et al.*, 2006
Fischerella ambigua 108b	*Bacillus cereus, Staphylococcus epidermidis, Pseudomonas aeruginosa, Escherichia coli, Micrococcus luteus.*	NI	Jaki *et al.*, 1999
Fischerella ambigua FS2	*Staphylococcus epidermidis, Escherichia coli.*	NI	Soltani *et al.*, 2005
Fischerella ambigua FS18	*Bacillus subtilis, Staphylococcus epidermidis, Escherichia coli, Candida kefyr.*	NI	Soltani *et al.*, 2005
Gloeocapsa caldariorum 127	*Bacillus subtilis*	NI	Mundt *et al.*, 2001

Contd...

Table Contd...

Hapalosiphon aureus FS3	*Bacillus subtilis, Staphylococcus epidermidis, Candida kefyr.*	NI	Soltani *et al.*, 2005
Hapalosiphon brasiliensis FS6	*Staphylococcus epidermidis*	NI	Soltani *et al.*, 2005
Hapalosiphon fontinalis FS7	*Bacillus subtilis, Staphylococcus epidermidis, Escherichia coli.*	NI	Soltani *et al.*, 2005
Hapalosiphon hibernicus	*Aspergillus oryzae, Candida albicans, Penicillium notatum, Saccharomyces cerevisiae, Trichophyton mentagrophytes*	Ambiguines	Smitka *et al.*, 1992
Hapalosiphon hibernicus FS33	*Bacillus subtilis, Staphylococcus epidermidis, Escherichia coli, Salmonella typhi, Candida kefyr, Candida albicans.*	NI	Soltani *et al.*, 2005
Hormothamnion enteromorphoides	*Bacillus subtilis, Escherichia coli, Pseudomonas aeruginosa, Salmonella typhimurium, Staphylococcus aureus, Candida albicans, Trichophyton mentagrophytes*	Hormothamnin A	Gerwick *et al.*, 1989
Limnothrix sp.	*Staphylococcus aureus, Bacillus subtilis, Micrococcus flavus.*	NI	Kreitlow *et al.*, 1999
Limnothrix redekei HUB 052	*Bacillus subtilis, Staphylococcus aureus, Micrococcus flavus*	α-linolenic acid	Mundt *et al.*, 2001
Lyngbya majuscula	*Mycobacterium smegmatis, Streptococcus pyogenes*	Malyngolide	Cardllina and Moore, 1979

Contd...

Table Contd...

Lyngbya majuscula	*Bacillus subtilis, Bacillus typhosus, Mycobacterium balnei, Mycobacterium phlei, Mycobacterium smegmatis, Sarcina lutea, Staphylococcus aureus, Candida albicans, Cryptococcus neoformans, Penicillum sp.*	NI	Moikeha and Chu (1971); Welch (1962); Gupta and Shrivastava (1965)
Lyngbya majuscula	*Candida albicans*	Tanikolide	Singh *et al.*, 1999
Lyngbya sp. 91/50	*Bacillus cereus*	NI	Jaki *et al.*, 1999
Lyngbya sp. 163	*Escherichia coli*	NI	Jaki *et al.*, 1999
Lyngbya sp. 164	*Bacillus cereus, Staphylococcus epidermidis.*	NI	Jaki *et al.*, 1999
Mastigocoleus testarum	*Escherchia coli, Mycobacterium phlei, Salmonella sp.*	NI	Kobbia and zaki, 1976
Mastigocladus laminosus	*Staphylococcus aureus, Candida albicans*	NI	Fish and Codd, 1994
Microcystis aeruginosa	*Bacillus subtilis, Bacillus cereus*	NI	Stensvik *et al.*, 1998
Nostoc commune	*Aspergillus candidus*	Nostofungicidine	Kajiyama *et al.*, 1998
Nostoc commune 122b	*Staphylococcus epidermidis*	NI	Jaki *et al.*, 1999
Nostoc commune 123b	*Bacillus cereus*	NI	Jaki *et al.*, 1999
Nostoc commune 124b	*Staphylococcus epidermidis*	NI	Jaki *et al.*, 1999
Nostoc commune 259	*Staphylococcus epidermidis*	NI	Jaki *et al.*, 1999
Nostoc commune (EAWAG 122B)	*Staphylococcus epidermidis, Bacillus cereus.*	NI	Jaki *et al.*, 2000

Contd...

Table Contd...

Nostoc commune NC1	*Bacillus subtilis, Micrococcus luteus, E. coli*	NI	Falch *et al.*, 1995
Nostoc linckia FS8	*Staphylococcus epidermidis, Escherichia coli.*	NI	Soltani *et al.*, 2005
Nostoc muscorum	*Staphylococcus aureus, Pseudomonas aeruginosa Candida albicans, Cunninghamella blakesleeana, Rhizoctonia solani, Sclerrotinia sclerotiorum*	NI	de Cano *et al.* (1990), Bloor and England (1989)
Nostoc muscorum 1453/12a	*Bacillus subtilis*	NI	Cannell *et al.*, 1988
Nostoc muscorum FS13	*Bacillus subtilis, Staphylococcus epidermidis, E. coli.*	NI	Soltani *et al.*, 2005
Nostoc punctiforme FS5	*Bacillus subtilis, Salmonella typhi*	NI	Soltani *et al.*, 2005
Nostoc sp.	*Escherichia coli, Staphylococcus epidermidis, Agrobacterium vitis, Penicillium expansum* and *Rhizoctonia solani*	NI	Piccardi *et al.*, 2000
Nostoc sp. FS4	*Bacillus subtilis*	NI	Soltani *et al.*, 2005
Nostoc sp. 51	*Bacillus subtilis, Micrococcus luteus, Penicillium oxalicum*	NI	Falch *et al.*, 1995
Nostoc spongiaeforme	*Bacillus subtilis, Staphylococcus aureus*	Tenuecyclamides A-D, Cyclic Hexapeptides	Banker and Carmeli, 1998
Nostoc species 104	*Bacillus subtilis*	NI	Mundt *et al.*, 2001

Contd...

Table Contd...

Oscillatoria angusta FS19	*Bacillus subtilis, Staphylococcus epidermidis, Candida kefyr.*	NI	Soltani *et al.*, 2005
Oscillatoria tenuis	*Staphylococcus aureus, Bacillus subtilis, Micrococcus flavus*	NI	Kreitlow *et al.*, 1999
Oscillatoria rubescens	*Staphylococcus aureus, Bacillus subtilis, Micrococcus flavus*	NI	Kreitlow *et al.*, 1999
Oscillatoria rubescens R19	Bacillus subtilis, *Micrococcus luteus, Escherichia coli, Penicillium oxalicum*	NI	Falch *et al.*, 1995
Oscillatoria species 234	*Bacillus subtilis*	NI	Mundt *et al.*, 2001
Oscillatoria sp.	*Staphylococcus aureus, Bacillus subtilis, Micrococcus flavus* *Pseudomonas aeruginosa, Saccharomyces cerevisiae*	NI	Kreitlow *et al.*, 1999 and Katircioglu *et al.*, 2006
Oscillatoria sp. 1459/8	*Bacillus subtilis*	NI	Cannell *et al.*, 1988
Oscillatoria redekei 1459/12a	*Bacillus subtilis*	NI	Cannell *et al.*, 1988
Oscillatoria princes	*Escherichia coli, Staphylococcus aureus, Bacillus subtilis, Bacillus typhosus, E. coli*	NI	Gupta and Srivastava, 1965
Phormidium sp.	*Staphylococcus aureus, Bacillus cereus, B. coagulans, B. licheniformis, B. sphaericus, B. subtilis, Streptococcus faecalis, S. viridi, E. coli, Proteus vulgaris, P. morganii, Pseudomonas denitrificans, Candida albicans*	NI	Fish and Codd, 1994

Contd...

Table Contd...

Phormedium sp. HS-273	*Staphylococcus aureus*	NI	Ohta *et al.*, 1993
Phormidium favosum 261	*Staphylococcus epidermidis*	NI	Jaki *et al.*, 1999
Phormidium sp. 178	*Candida albicans*	NI	Jaki *et al.*, 1999
Phormidium sp. 211	*Candida albicans*	NI	Jaki *et al.*, 1999
Plectonema notatum FS28	*Bacillus subtilis*	NI	Soltani *et al.*, 2005
Pseudoanabaena catenata	*Bacillus subtilis*	NI	Mundt *et al.*, 2001
Pseudoanabaena catenata U7	*Bacillus subtilis, Micrococcus luteus, Penicillium Oxalicum*	NI	Falch *et al.*, 1995
Spirulina platensis	*Streptococcus faecalis, Bacillus subtilis, Staphylococcus aureus, S. epedermidis, Enterobacter aerogenes, E. cloacae, Escherichia coli, Pseudomonas aeruginosa, Salmonella typhimurium, Proteus vulgaris, Candida albicans*	Heptadecane	Ozdemir *et al.* 2004
Schizothrix sp. 1470/2	*Staphylococcus aureus*	NI	Cannell *et al.*, 1988
Synechococcus leopoliensis	*Bacillus subtilis*	NI	Cannell *et al.*, 1988
Scytonema lyngbyoides	*Bacillus subtilis, Micrococcus luteus, Escherichia coli*	NI	Falch *et al.*, 1995
Scytonema myochrous 200b	*Staphylococcus epidermidis*	NI	Jaki *et al.*, 1999

Contd...

Table Contd...

Stigonema ocellatum FS23, FS 30 & FS 14	*Bacillus subtilis, Staphylococcus epidermidis, E. coli, Candida keefyr*	NI	Soltani *et al.*, 2005
Symploca muscorum 176	*Micrococcus luteus, Candida albicans*	NI	Jaki *et al.*, 1999
Scytonema sp.	*Aspergillus oryzae, Candida albicans, Penicillium notatum, Saccharomyces cerevisiae*	Tolytoxin and 6-Hydroxy-7-O-Mehtylescytophycine	Carmeli *et al.*, 1990
Tolypothrix tjipanensis	*Aspergillus oryzae, Candida albicans, Trichophyton mentagrophytes*	NI	Bonjouklian *et al.*, 1991
Tolypothrix byssoidea 195	*Micrococcus luteus, Candida albicans*	NI	Jaki *et al.*, 1999
Tolypothrix byssoidea (EAWAG195)	*Candida albicans*	Tolybyssidin A (1) Tolybyssidin B (2)	Jaki *et al.*, 2001
Tolypothrix distorta 153a	*Candida albicans*	NI	Jaki *et al.*, 1999
Tolypothrix distorta 96c	*Candida albicans*	NI	Jaki *et al.*, 1999
Tychonema bourrellyi	*Bacillus subtilis, Bacillus cereus, Escherichia coli*	NI	Stensvik *et al.*, 1998
Westiellopsis prolifica	*Aspergillus oryzae, Candida albicans, Penicillium notatum, Saccharomyces cerevisiae, Trichophyton mentagrophytes*	Ambiguines	Smitka *et al*, 1992

NI-Not Identified

Tuney *et al*. (2006) tested the methanol, acetone, diethyl ether and ethanol extracts of eleven sea-weed species from the coast of Urla *in vitro* for their antimicrobial activities against *Candida sp., Enterococcus faecalis, Staphylococcus aureus, Streptococcus epidermidis, Pseudomonas aeruginosa* and *Escherichia coli* with disc diffusion method. They have also reported that diethyl ether was the best solution for extracting the effective antimicrobial material from the algal species used in the experiment, with the exception of *D. linearis* (Marine algae) for which ethanol was the most effective extraction solution.

Rao *et al*. (2007) reported the antimycobacterial activity of Cyanobacteria (*Hapalosiphon sp., Anabaena sp. Lyngbya sp., Westeillopsis prolifica, Spirulina sp. Anabaena variabilis, Anabaena cylindrica, Oscillatoria sp.* and *Scytonema sp.*) against *Mycobacterium tuberculosis ATCC 27294, M.tuberculosis MDR, M.avium, M. intracellulare and M. aurum.*

Chapter 3

BACTERIOLOGY OF TEST MICROORGANISMS

Cyanobacterial species

Cyanobacterial species *viz., Anabaena variabilis, Anabaena fertilissima, Nostoc muscorum, N. punctiforme, N. linckia, N. commune, Spirulina platensis, Westiellopsis prolifica* and *Hapalosiphon sp.* were obtained from National Center for Conservation and Utilization of Blue-green Algae, Indian Agricultural Research Institute, New Delhi. Cultures were maintained in BG-11 growth medium throughout the study in the laboratory.

Systematic Position : Nomenclature and classification of selected species of cyanobacteria by Holt *et al.* (1994) has been followed here.

Anabaena

Kingdom	Bacteria
Division	Cyanophyta
Class	Cyanophyceae
Order	Nostocales
Family	Nostocaceae
Genus	*Anabaena*
Species	*Anabaena variabilis*
	Anabaena fertilissima

General Characteristics : Trichomes are untapered with conspicuous constrictions at cross-walls, trichomes may be straight, curved or helically formed, the cells are cylindrical, spherical or ovoid and not shorter than broad, usually ranging in width from 2-10 ìm but in some species to over 20 µm (Desikachary, 1959). The terminal cells may be rounded, tapered or conical in shape, *heterocysts* are intercalary or terminal or both, *akinetes* are usually formed, and their position in trichomes differs with species. A firm individual sheath is absent, but a soft mucilaginous covering is often present, trichomes, when free of adhesive mucilage, are normally motile, and colonies are not formed, reproduction is by fragmentation of parental trichome. Many species are known worldwide as major components of the freshwater plankton and also of many saline lakes. Generally within a mucilaginous sheath, *hormogonia* arise from such aggregates, asymmetric division of vegetative cells occurs, small heterocysts develop when strains are deprived of combined nitrogen, masses of vegetative cells may also enlarge to form thick walled cells, akinete germination takes place with division in several planes and shedding of the extra wall layers.

Nostoc

Kingdom Bacteria
Division Cyanophyta
Class Cyanophyceae
Order Nostocales
Family Nostocaceae
Genus *Nostoc*
Species *Nostoc muscorum*
Nostoc punctiforme
Nostoc linckia
Nostoc commune

General Characteristics : The trichomes are untapered with conspicuous constrictions at cross-walls, cells are cylindrical, spherical or ovoid and are not shorter than broad, *heterocysts* are intercalary under most circumstances, *Nostoc* is characterized by a confluent gel holding masses of trichomes together, often in the form of a massive thallus which may be spherical, ovoid or of a less discernible shape (Vaucher, 1803). *Nostoc* is distinguished by the presence of a

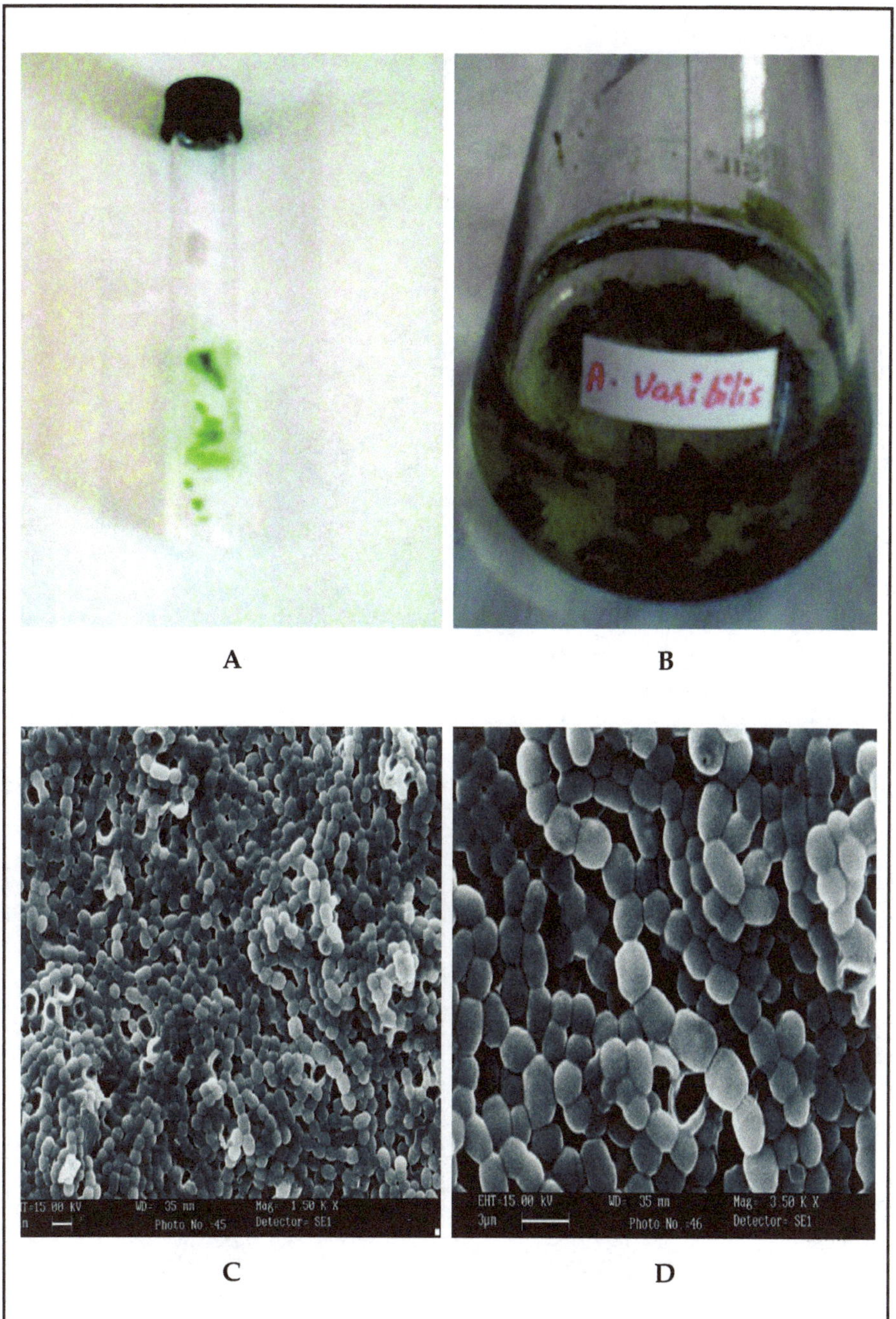

Fig. 4. Different features of *Anabaena variabilis* : **A.** Growth on BG-11 agar medium. **B.** Twenty five days old culture. **C. & D.** Scanning electron micrographs at different magnifications.

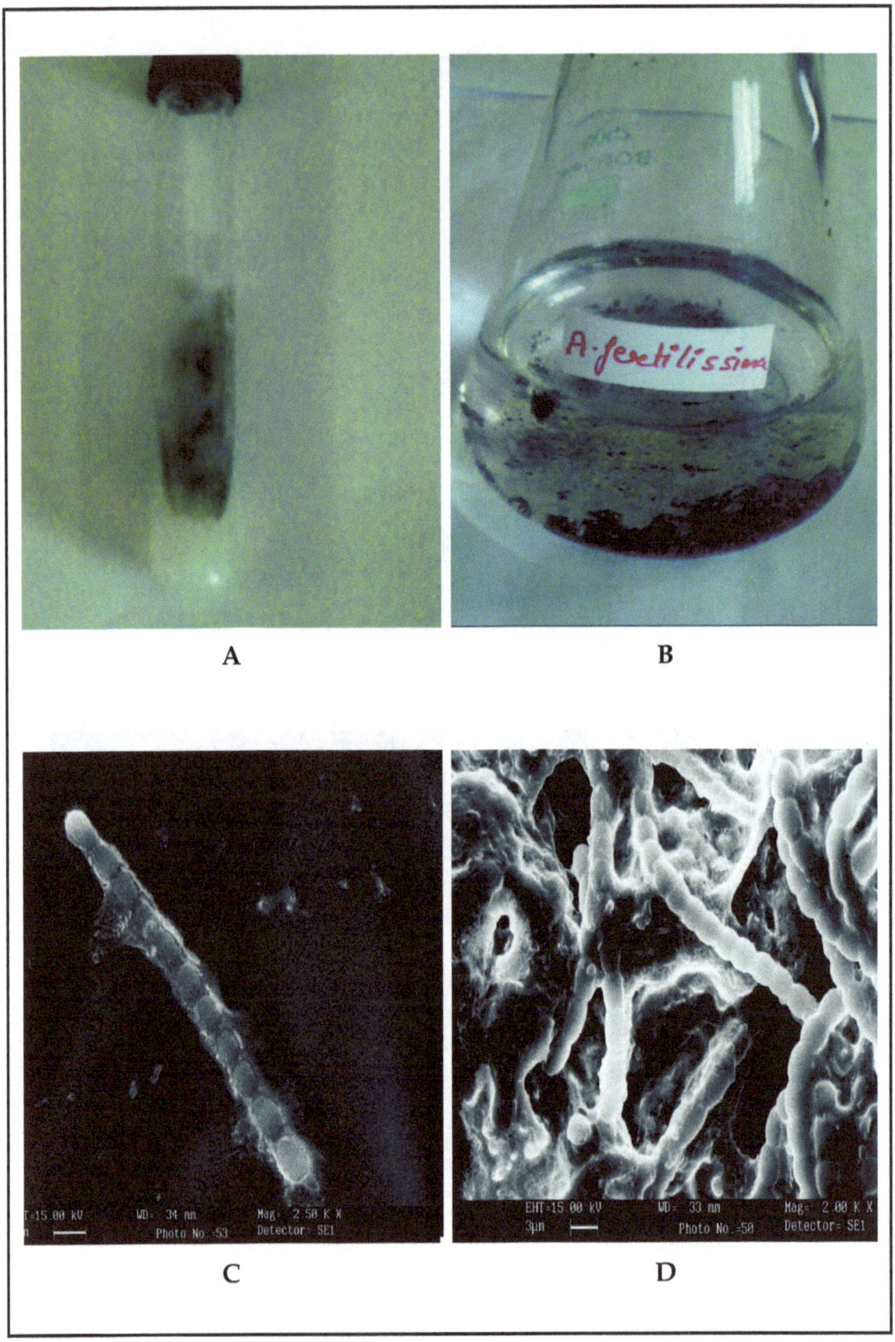

A B C D

Fig. 5. Different features of *Anabaena fertilissima* : **A.** Growth on BG-11 agar medium. **B.** Twenty five days old culture. **C. & D.** Scanning electron micrographs at different magnifications.

Fig. 6. Different features of *Nostoc muscorum* : **A.** Growth on BG-11 agar medium. **B.** Twenty five days old culture. **C. & D.** Scanning electron micrographs at different magnifications. Some cells occasionally show outgrowth on one side at right angles to the length of the trichome.

Fig. 7. Different features of *Nostoc punctiforme* : **A.** Growth on BG-11 agar medium. **B.** Twenty five days old culture. **C. & D.** Scanning electron micrographs at different magnifications.

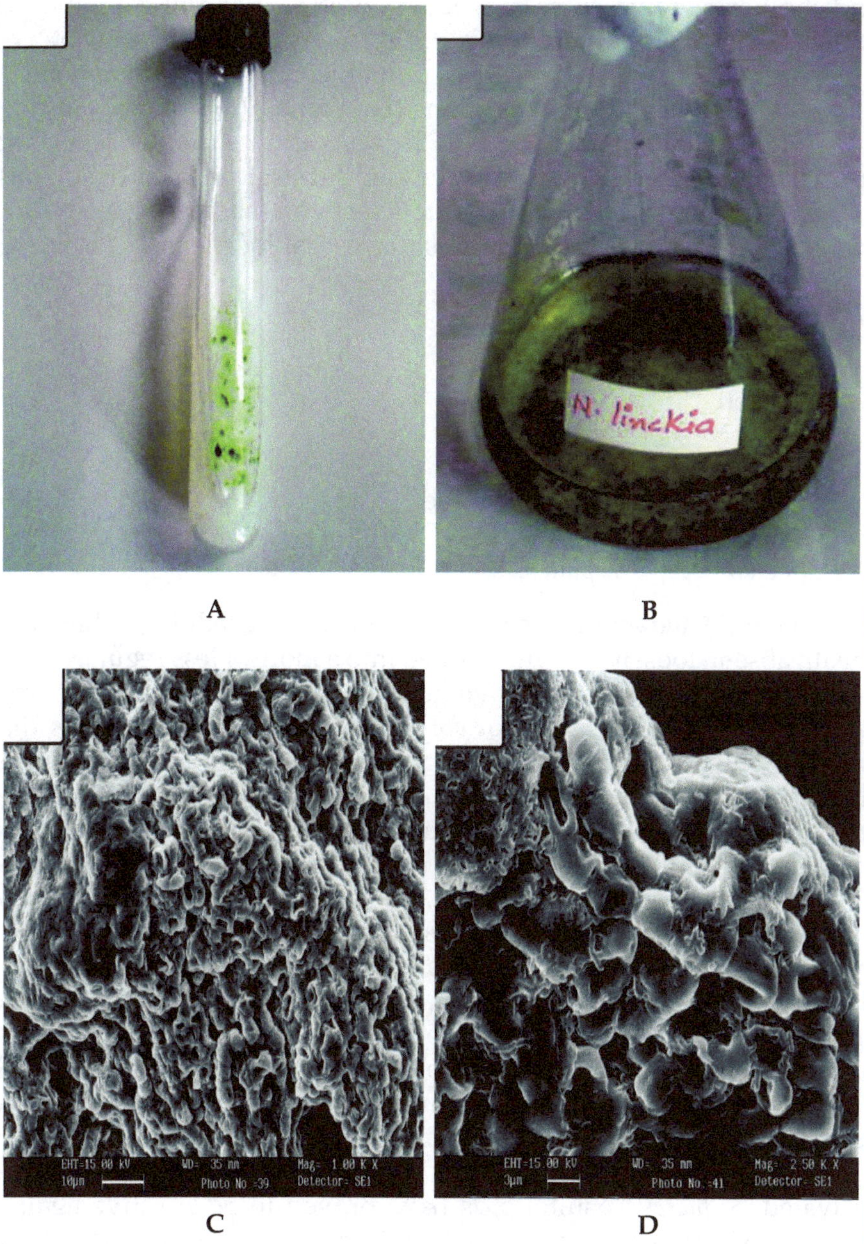

Fig. 8. Different features of *Nostoc linckia*: **A.** Growth on BG-11 agar medium. **B.** Twenty five days old culture. **C. & D.** Scanning electron micrographs at different magnifications.

developmental cycle, short chains of cells are formed and released, and the hormogonia cease movement after a period of migration at which time terminal heterocysts are differentiated, followed by cell division and growth which is accompanied by *gel formation* under natural and some culture conditions (Lazaroff, 1973). Various species of *Nostoc* are known from benthic habitats in fresh water lakes and springs, some species are attached to solid substrate in lakes or streams, many species occur as amorphous sheets or masses of gel-bound trichomes in fresh water or in moist terrestrial locations.

Spirulina

Kingdom	Bacteria
Division	Cyanophyta
Class	Cyanophyceae
Order	Oscillatoriales
Family	Oscillatoriaceae
Genus	*Spirulina*
Species	*S. platensis*

General Characteristics : Trichomes are multicellular, cylindrical, sheath absent, loosely or tightly coiled into a more or less regular spiral, apex or trichome usually not attenuated, cross walls if present obscured, terminal cell rounded, without calyptra, filamentous organisms that divide exclusively by binary fission and in one plane but that grow in the form of tight to nearly tight coiled right or left handed helix, gliding motility consists of a turning of the screw thus, with great transverse movement and little forward motion. The terminus of the trichome is either blunt or pointed in different species, the width of the trichome may be from 1-5 m, and colour is variable, blue green to red. The cyanobacterium *Spirulina platensis* exhibits a diverse biological activity due to high content of highly valuable proteins, indispensable amino acids, vitamins, beta-carotene and other pigments, mineral substances, indispensable fatty acids and polysaccharides (Allnutt, 1996). *S. platensis* produces an immunostimulating effect by enhancing the resistance in humans, mammals, chickens and fish against infections due to the capacity of *S. platensis* macrophages thus B cells are activated. *S. platensis* sulfolipids have proved to be effective against HIV. Preparation obtained from *S. platensis* biomass has also been found active against herpesvirus, cytomegalovirus, influenza virus, etc. *S. platensis* extracts are capable in inhibiting carcinogenesis due to anti-oxidant properties that protect tissues and also reduce toxicity in liver kidney and tests (Khan *et al.*, 2005).

Fig. 9. Different features of *Nostoc commune* : **A.** Growth on BG-11 agar medium. **B.** Twenty five days old culture. **C. & D.** Scanning electron micrographs at different magnifications.

Westiellopsis

Kingdom	Bacteria
Division	Cyanophyta
Class	Cyanophyceae
Order	Stigonematales
Family	Stigonemataceae
Genus	*Westiellopsis*
Species	*W. prolifica*

General Characteristics : Thallus filamentous with true branching-filaments of two kinds, primary filament slightly thicker and more or less creeping, secondary filaments generally thinner and growing erect, filaments without a sheath and consisting of a single row of cells, heterocyst intercalary.

Hapalosiphon

Kingdom	Bacteria
Division	Cyanophyta
Class	Cyanophyceae
Order	Stigonematales
Family	Stigonemataceae
Genus	*Hapalosiphon*

General Characteristics : Thallus caespitose, floccose, thin, aquatic, filaments free, not coalescing laterally, cells in one or two rows, sheath present, continuously branched; branches irregularly lateral, true, often arising only on one side of the filaments, false branches present, branches erect from the primary prostrate filaments, erect branches as broad as and similar to the main filament, heterocysts intercalary, only occasionally lateral, hormogonia formed mostly from the side branches, spores present.

Test Micro-organisms

Test micro-organisms used in the present study were collected from Institute of Microbial Technology (IMTECH), Chandigarh, India. The species employed include pathogenic Gram-positive and Gram-negative bacteria as listed below:

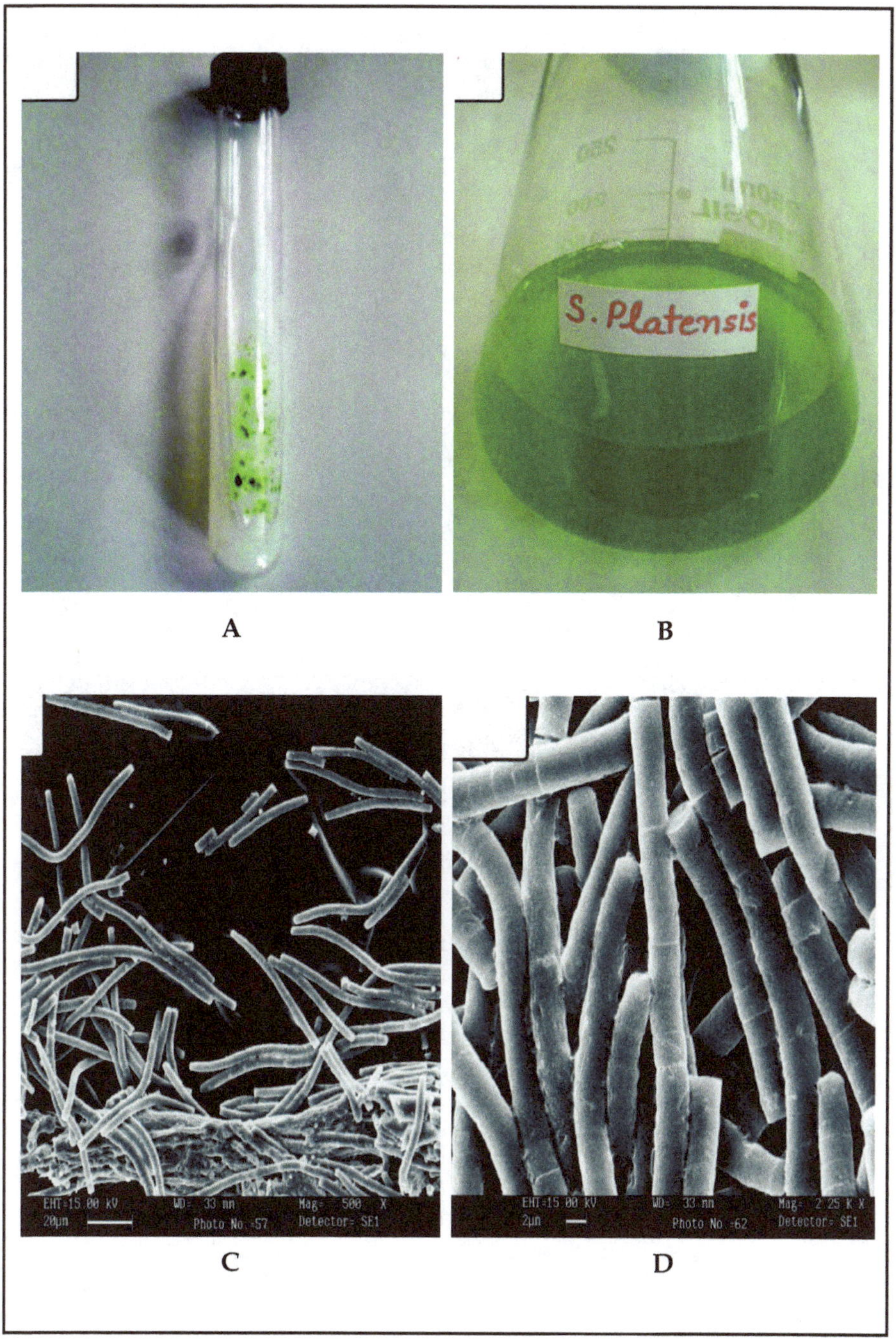

Fig. 10. Different features of *Spirulina platensis*: **A.** Growth on BG-11 agar medium. **B.** Twenty five days old culture. **C. & D.** Scanning electron micrographs at different magnifications.

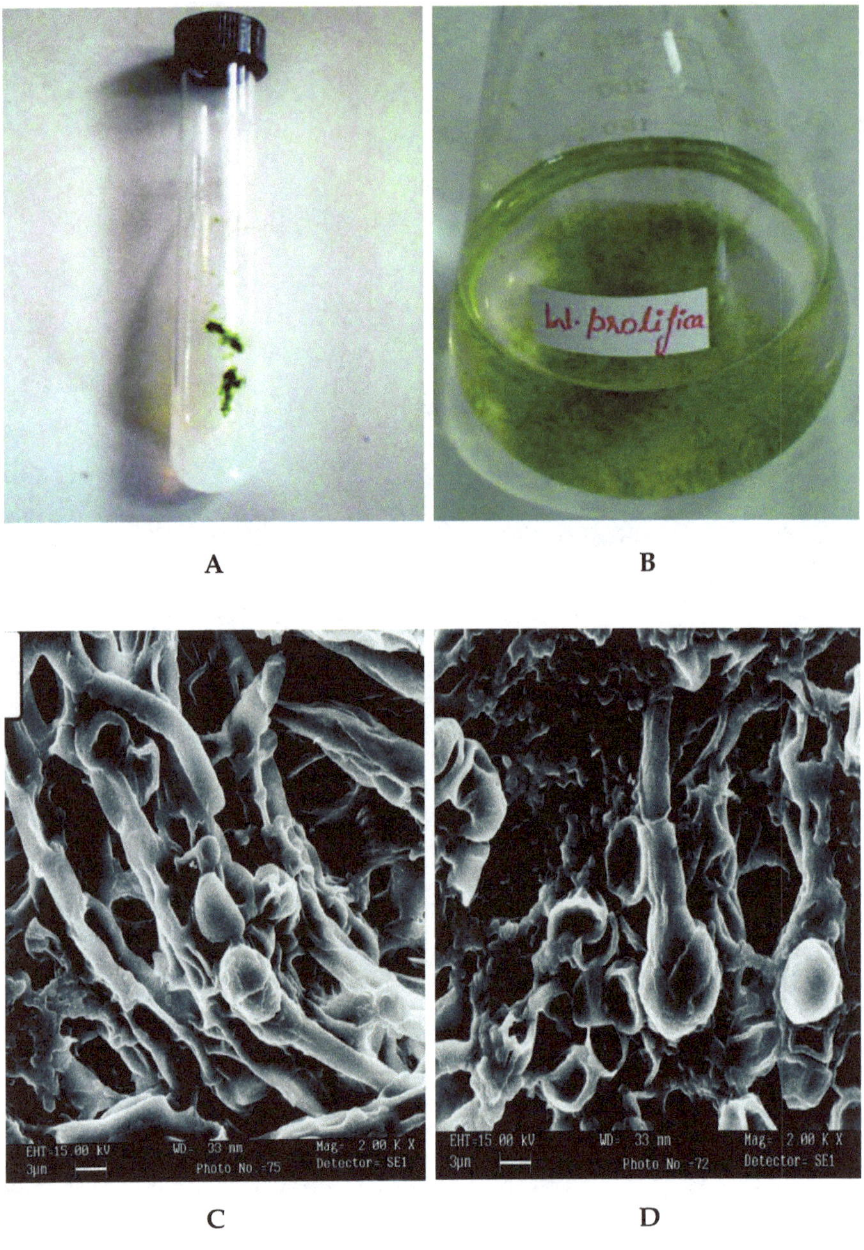

Fig. 11. Different features of *Westiellopsis prolifica*: **A.** Growth on BG-11 agar medium. **B.** Twenty five days old culture. **C. & D.** Scanning electron micrographs at different magnifications.

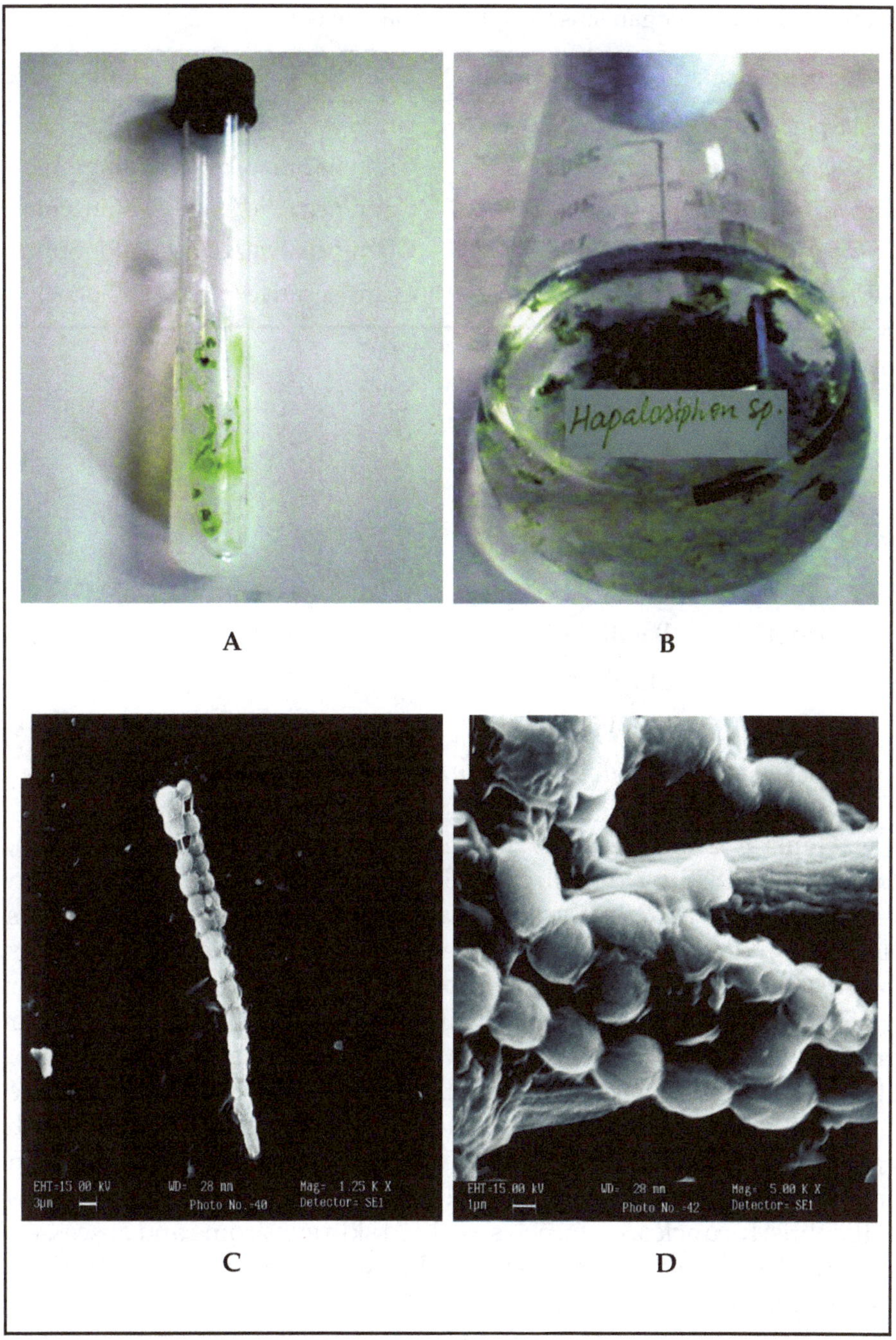

Fig. 12. Different features of *Hapalosiphon* sp.: **A.** Growth on BG-11 agar medium. **B.** Twenty five days old culture. **C. & D.** Scanning electron micrographs at different magnifications.

Table 6. Test microorganisms for antibacterial assay

Microorganisms	Source	Gram Stain/Shape	Nature
Staphylococcus aureus	MTCC 740	Gram-Positive/cocci	Pathogenic
Pseudomonas aeruginosa	MTCC 741	Gram-negative/rods	Pathogenic
Escherichia coli	MTCC 723	Gram-negative/rods	Pathogenic
Salmonella typhi	MTCC 733	Gram-negative/rods	Pathogenic
Klebsiella pneumoniae	MTCC 139	Gram-negative/rods	Pathogenic

Staphylococcus aureus

Domain	Bacteria
Kingdom	Eubacteria
Phylum	Firmicutes
Class	Bacilli
Order	Bacillales
Family	Staphylococcaceae
Genus	*Staphylococcus*
Species	*Staphylococcus aureus*

General Characteristics : *S. aureus* was discovered in Aberdeen, Scotland in 1880 by the surgeon Sir Alexander Ogston in pus from surgical abscesses. *S. aureus* is a Gram-positive coccus, which appears as grape-like clusters when viewed through a microscope and has black, shiny, colonies surrounded by clear zones on Baird parker agar plate and yellow colonies with yellow zones on Mannitol salt agar plate (Ryan and Ray, 2004). *S. aureus* is also known as golden bacteria. It is a spherical bacterium, frequently living on the skin or in the nose of a person. Approximately 20–30% of the human population is "staph carriers" (Heyman, 2004). *S. aureus* can cause a range of illnesses from minor skin infections, such as pimples, impetigo, boils, cellulitis folliculitis, furuncles, carbuncles, scalded skin syndrome and abscesses, to life-threatening diseases, such as pneumonia, meningitis, osteomyelitis endocarditis.

S. aureus can grow at a temperature range of 15 to 45°C and at NaCl concentrations as high as 15%. Nearly all strains of *S. aureus* produce the enzyme coagulase

Macroscopic Examination

Characteristics colonies grow on specific type of media which are as follows

Table 7 : Morphological characteristics of *Staphylococcus aureus* on selective agar media

Selective Medium	Characteristic Colony Morphology	Gram Stain
Baird-Parker Agar Medium (BPA)	Black, shiny, surrounded by clear zones 2 to 5mm	Positive cocci (in clusters)
Mannitol-Salt Agar Medium (MSA)	Yellow colonies with yellow zones	Positive cocci (in clusters)
Vogel-Johnson Agar Medium (VJA)	Black surrounded by yellow zone	Positive cocci (in clusters)

Biochemical Identification

Gram's Stain

Principle

This stain is mainly used to differentiate between Gram positive and negative bacteria. It also helps in identifying the bacterial morphology.

Procedure

- Pick the colony from the nutrient agar slant and place it on a clean grease free slide.
- Prepare a thin smear by addition of sterile water in it.
- Allow it to air dry and fix by gentle heat.
- Cover the smear with crystal violet solution for 1 minute, pour off the stain and wash with tap water.
- Cover the slide with iodine solution and allow remaining for 1 minute.
- After 1 minute, wash the slide first with decolourizer until the dye ceases to stream out.
- Wash the slide in running tap water and dip in safranin solution for about one minute.
- Wash the slide in tap water and dry.
- After drying examine under microscope.

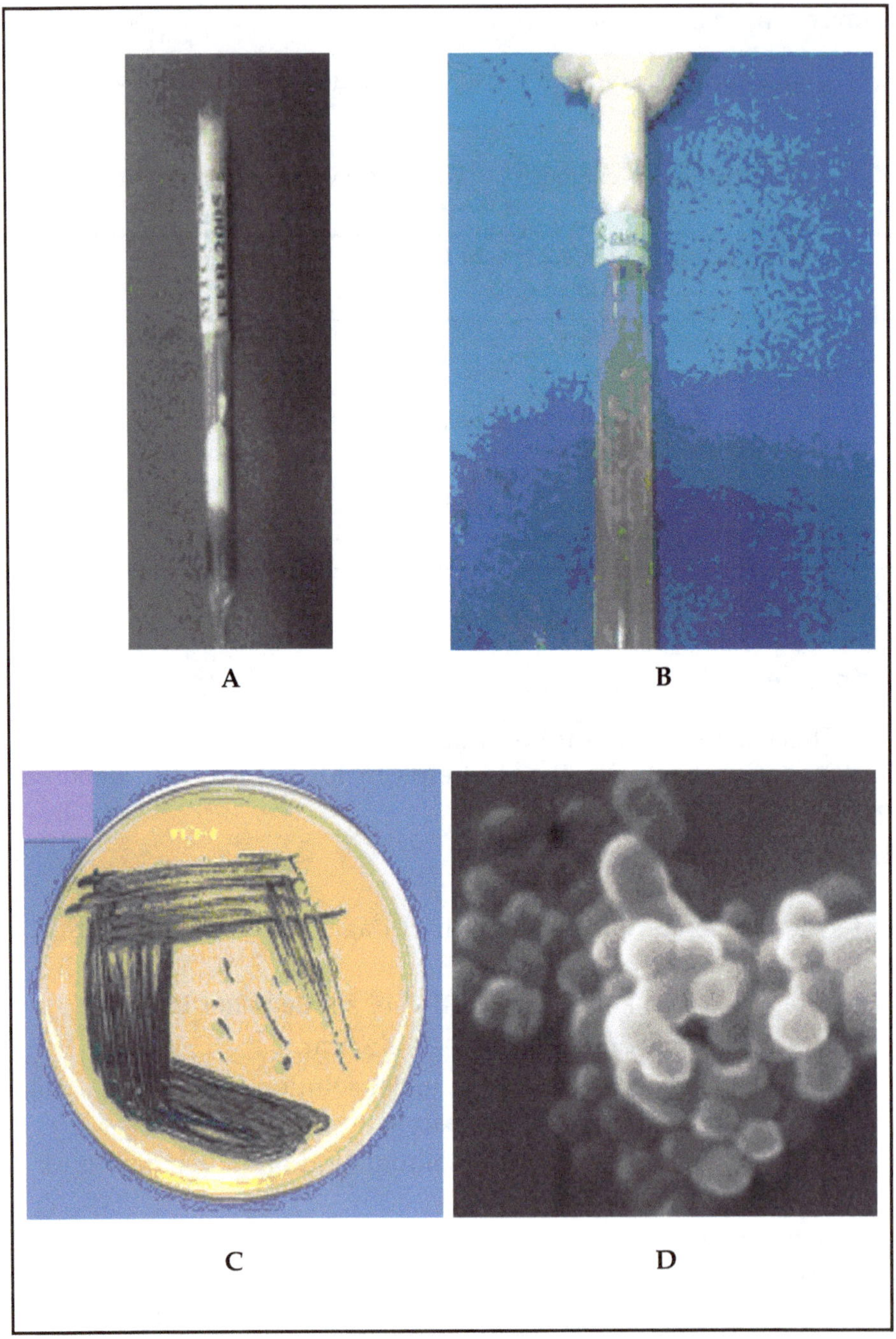

Fig. 13. Different features of *Staphylococcus aureus* : **A.** Lyophilized vial. **B.** Growth on nutrient agar slant. **C.** Characteristic colonies on baird parker agar plate. **D.** Scanning electron micrograph at 10 KX magnification.

Coagulase Test
Principle

This test is based on the bacteria's ability to clot blood plasma using the enzyme coagulase. If the organism has coagulase it will clump rabbit plasma. *S. aureus* produces two types of coagulase, free and bound. Free coagulase is an extracellular enzyme produced when the organism is cultured in broth. Bound coagulase, also known as the clumping factor, remains attached to the cell wall of the organism.

The tube test can detect the presence of both bound and free coagulase. Isolates that do not produce clumping factor must be tested for the ability to produce extracellular coagulase (free coagulase). In the direct tube test, free coagulase liberated from the cell acts on prothrombin in the coagulase plasma to give a thrombin-like product. This product then acts on fibrinogen to form a fibrin clot.

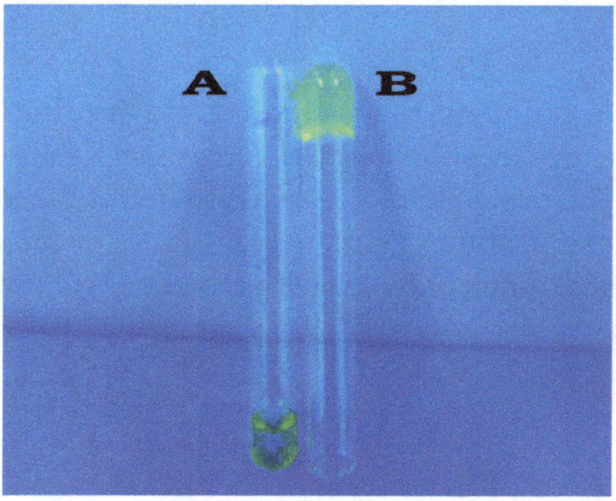

Fig. 14. Coagulase test: A, Control tube. B, Tube shows the clot formation, hence test is positive.

Procedure

- Transfer the selective colonies from the nutrient agar surface with the help of sterile inoculating loop into a tube of Sterile nutrient broth. Incubate the tube for 18-24 hrs or until a dense growth is observed.
- Using a sterile 1-ml pipette, add 500 μl of rabbit Coagulase Plasma (BBL).

- Using a sterile serological pipette, add approximately 50µl of the overnight broth culture of the test organism to the Coagulase plasma tube.
- Mix the tube content properly.
- Incubate in a water bath or in an incubator at 35 - 37°C for up to 4hrs.
- Examine the tubes periodically by gently tipping, without shaking or agitating the tubes at 3 hrs and subsequently at suitable interval up to 24 hrs.
- Examinine the tubes for the clot.
- Similarly positive and negative controls are also prepared simultaneously for comparison.

Catalase test

Principle

Catalase is found in most aerobic and facultative anaerobic bacteria. The main exception is *Streptococcus spp*. Catalase is not found in anaerobes. The principle of the test is to detect the presence of the enzyme catalase. Catalase enzyme is found in most bacteria. It catalyses the breakdown of hydrogen peroxide (H_2O_2) with the release of free oxygen.

Method

- Firstly dip a capillary tube into 3% H_2O_2
- After that touch a colony
- Observe the tube for bubble indicating a positive reaction

Biochemical Test Results

Test	Results
Gram-Staining	Gram-Positive
Catalase	Positive
Coagulase	Positive

Pseudomonas aeruginosa

Kingdom	Bacteria
Phylum	Proteobacteria
Class	Gamma Proteobacteria
Order	Pseudomonadales
Family	Pseudomonadaceae
Genus	*Pseudomonas*
Species	*Pseudomonas aeruginosa*

General Characteristics : The word *Pseudomonas* means 'false unit'. The stem word "mon" was used early in the history of microbiology to refer to germs, *e.g.* Kingdom Monera. The species name *aeruginosa* is derived from the Greek prefix "ae" meaning "old" or "aged" and the suffix "ruginosa" means 'wrinkled or bumpy' (Brown, 1956).

P. aeruginosa is member of the Gamma Proteobacteria class of Bacteria. It belongs to the bacterial family Pseudomonadaceae. *P. aeruginosa* is a Gram-negative, aerobic, rod-shaped bacterium measuring 0.5 to 0.8 μm by 1.5 to 3.0 μm. Almost all strains are motile by means of a single polar flagellum (Ryan and Ray, 2004).

Its optimum temperature for growth is 37°C, and it is able to grow at temperatures as high as 42°C. An opportunistic human pathogen *P. aeruginosa* secretes a variety of pigments, including pyocyanin, fluorescein and pyorubin. Although classified as an aerobic organism, *P. aeruginosa* is considered by many as a facultative anaerobe as it is well adapted to proliferate in conditions of partial or total oxygen depletion.

This organism can achieve anaerobic growth with nitrate as a terminal electron acceptor. *P. aeruginosa* is an opportunistic pathogen, meaning that it exploits some break in the host defenses to initiate an infection. It causes urinary tract infections, respiratory system infections, dermatitis, soft tissue infections, bacteremia, bone and joint infections, gastrointestinal infections and a variety of systemic infections, particularly in patients with severe burns and in cancer and AIDS patients who are immuno suppressed.

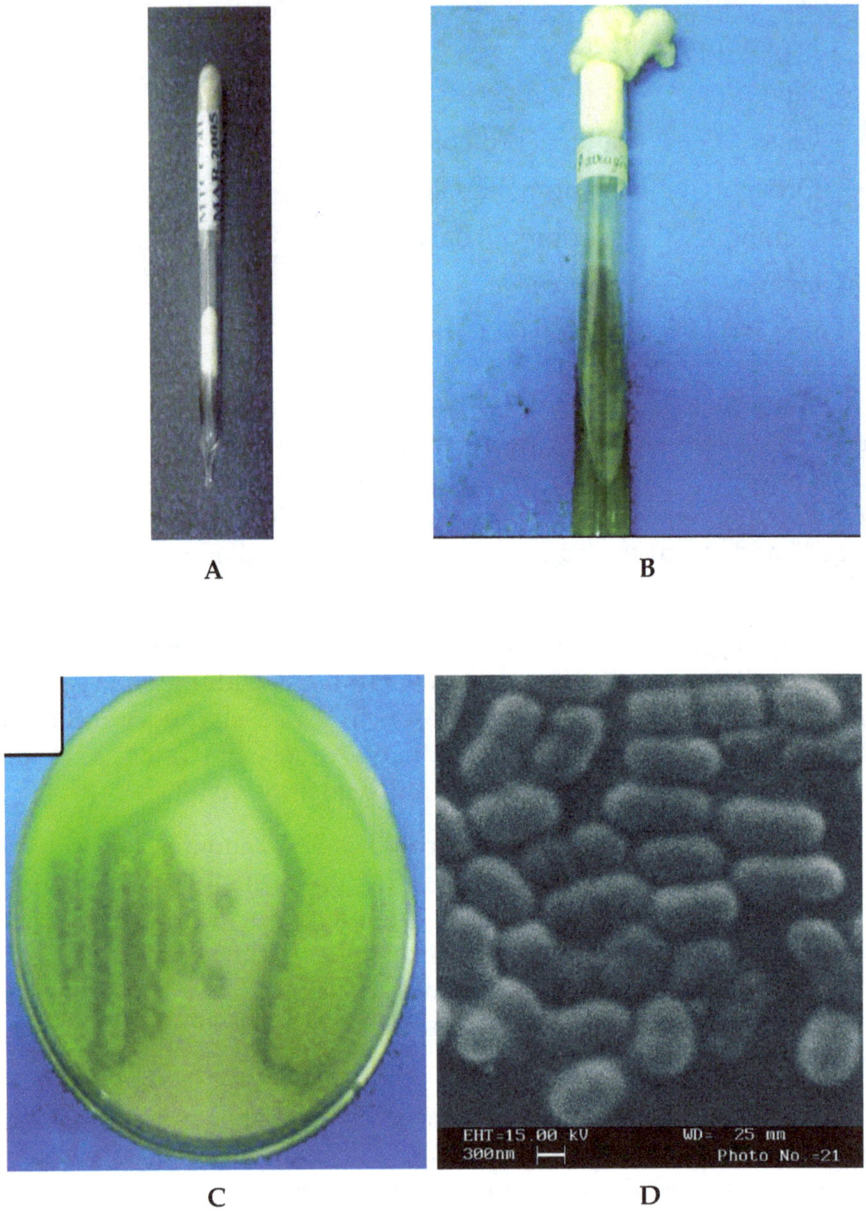

Fig. 15. Different features of *Pseudomonas aeruginosa*: **A.** Lyophilized vial. **B.** Growth on nutrient agar slant. **C.** Characteristic colonies on nutrient agar plate. **D.** Scanning electron micrographs at 12.00 KX magnification.

Macroscopic Examination

Characteristic colonies grow on specific type of media which are as follows.

Table 8. Morphological Characteristics of *Pseudomonas aeruginosa* on Agar Media

Selective Medium	Characteristic Colony Morphology	Gram Stain
Cetramide Agar Medium	Generally greenish	Negative rods
Nutrient agar	White colonies	Negative rods

Oxidase Test
Principle

To determine the presence of oxidase enzymes in the bacteria. The reagent (impregnated into strips of filter paper) contains tetramethyl-p-phenylenediamine dihydrochloride that serves as an alternative substrate for the cytochrome oxidase reaction. In the reduced state the reagent is colourless but when oxidised it becomes purple.

Method

Transfer the colonies from the nutrient agar plates onto strips or disks of filter paper (pink colour) that previously has been impregnated with N-diethyl-p-phenylenediamine hydrochloride.

Nitrate Reduction Test
Principle

To determine the ability of an organism to reduce nitrate to nitrites or free nitrogen gas

Method

- Inoculate a nitrate broth and incubate for 5 days at 37°C.
- To each nitrate broth culture, add 1 ml of sulphanilic acid and 1 ml of alpha-naphtalamine. The production of a red colour occurs in the presence of nitrite and indicates the ability of the organism to reduce nitrate to nitrite.

- To broth showing a negative reaction add a few particles of zinc. The appearance of a red colour indicates that nitrate is still present and hence has not been reduced by the organism. If the solution does not change colour the organism has reduced the nitrate through nitrite to nitrogen gas.

Hugh and Leifson (H-L) test

Principle

This test is used to determine whether a bacterium has the enzymes necessary for the aerobic breakdown of glucose (*i.e.* oxidation) and/or for the fermentation of glucose.

Method

- Stab a loopful of the culture from nutrient agar plates onto two tubes of H-L agar medium to within 1 cm of the bottom of the tube.
- Pour approx. 1 ml sterile liquid paraffin above the H-L agar on one tube only.
- Plug both the tubes with cotton and incubate at 37°C for 48 h.

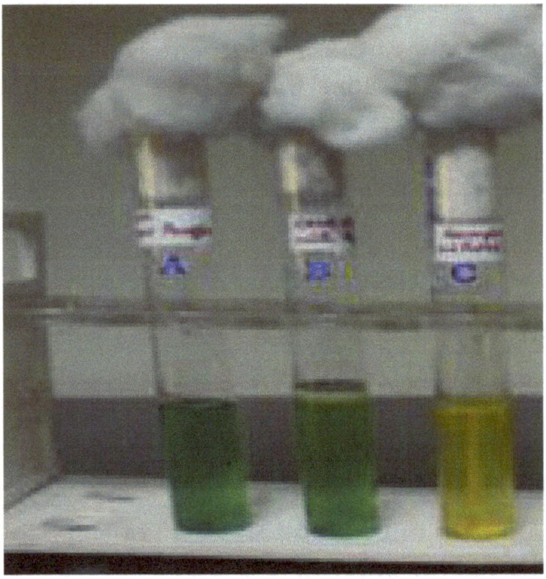

Fig. 16. Hugh and Leifson test: **A.** Media control tube. **B.** Tube having the liquid paraffin (no growth). **C.** Tube shows the yellow growth (Oxidative).

Observation

Open	Closed	Result
Yellow	Green	*Pseudomonas* present (Oxidation occurs but fermentation does not occur)
Yellow	Yellow	*Pseudomonas* absent (Oxidation does not occur but fermentation occurs)
Green	Green	*Pseudomonas* absent (Oxidation does not occur but fermentation occurs)

Biochemical Test Results

Test	Results
Gram-Staining	Gram-Positive
Catalase	Positive
Oxidase	Positive
Hugh-Leifson	Oxidative
Growth at 42°C	Positive
at 4 °C	Negative
Nitrate Reduction	Positive
Casein Hydrolysis	Positive

Escherichia coli

Kingdom	Bacteria
Phylum	Proteobacteria
Class	Gamma Proteobacteria
Ordar	Enterobacteriales
Genus	*Escherichia*
Species	*E. coli*

General Characteristics : German pediatrician and bacteriologist Theodor Escherich discovered *E. coli* in 1885 (Feng *et al.*, 2002). *E. coli* is a facultative anaerobic, Gram-negative and non-sporulating organisms. It is now classified as part of the Enterobacteriaceae family of gamma-proteobacteria. It can live on a wide variety of substrates. *E. coli* uses mixed-acid fermentation in anaerobic conditions, producing lactate, succinate, ethanol, acetate and carbon dioxide. Optimal growth of *E. coli* occurs at 37°C, but some laboratory strains can multiply at temperatures of up to 49°C. Strains that possess flagella can swim and are motile, but other strains lack flagellum. The flagella of *E. coli* have a peritrichous arrangement (Fotadar *et al.*, 2005).

E. coli normally colonizes an infant's gastrointestinal tract within 40 h of birth, arriving with food or water or with the individuals handling the child. It is a bacterium that is commonly found in the lower intestine of warm-blooded animals. *E. coli* are not always confined to the intestine, and their ability to survive for brief periods outside the body makes them an ideal indicator organism to test environmental samples for faecal contamination.

Most *E. coli* strains are harmless, but some, can cause serious food poisoning in humans, and are occasionally responsible for costly product recalls. The harmless strains are part of the normal flora of the gut, and can benefit their hosts by producing vitamin K2, or by preventing the establishment of pathogenic bacteria within the intestine. Strains of *E. coli* can cause gastroenteritis, urinary tract infections, and neonatal meningitis. In rarer cases, virulent strains are also responsible for peritonitis, mastitis, septicaemia and Gram-negative pneumonia (Todar, 2007).

Macroscopic Examination

Characteristics colonies grow on specific type of media which are as follows

Table 9 : Morphological Characteristics of *Escherichia coli* on Selective Agar Media

Selective Medium	Characteristic Colony Morphology	Gram Stain
MacConkey Agar Medium (MCA)	Pink colonies	Gram-Negative rods
Eosin Methylene Blue (EMB)	Purple colonies with green metallic sheen	Gram-Negative rods
Rapid *E. coli* Agar (REC)	Purple colonies	Gram-Negative rods

Biochemical Reaction

Indole Reaction

Principle

This test is done to determine the ability of the bacteria that can breakdown the aminoacid tryptophan to indole. Bacteria use an

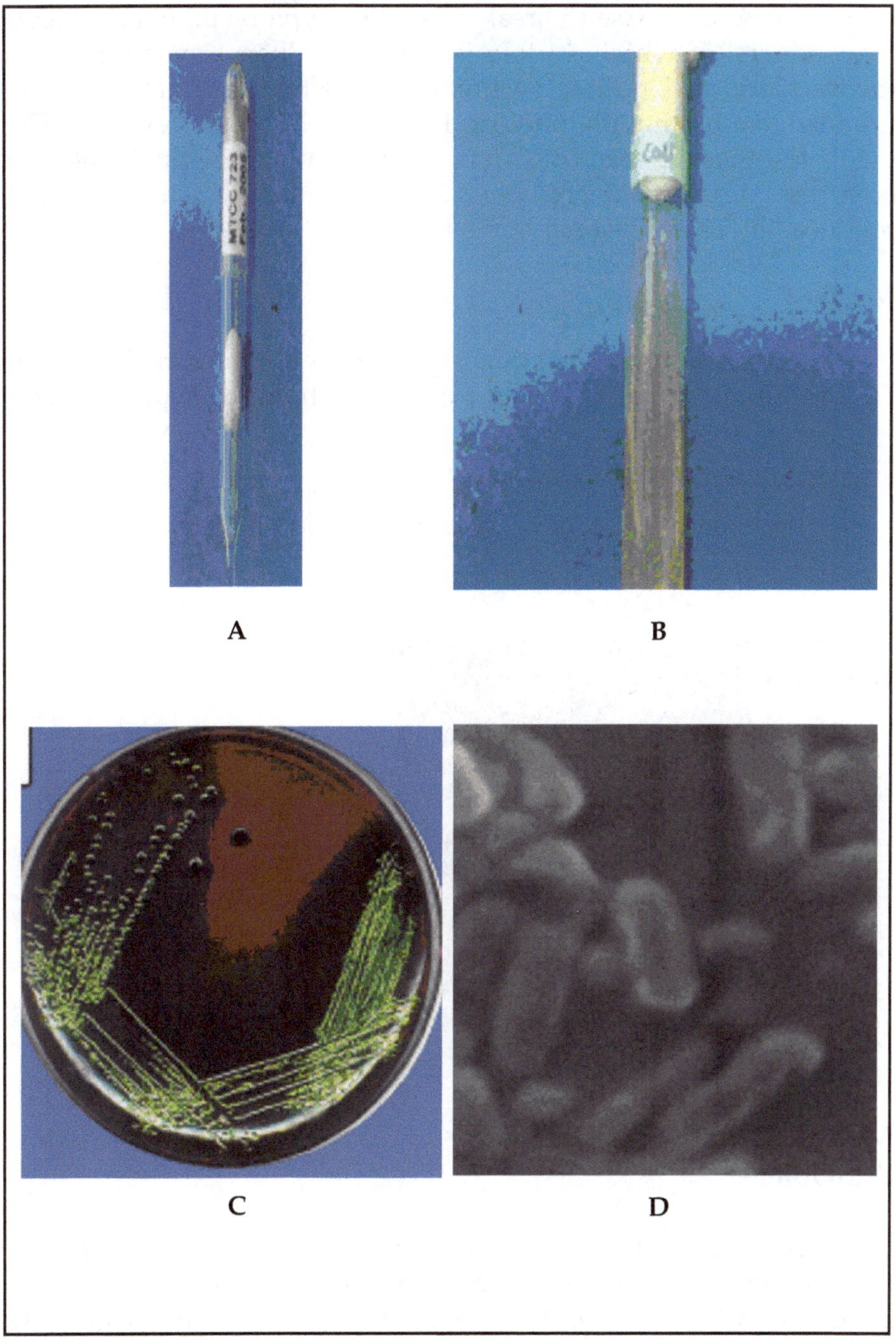

Fig. 17. Different features of *Escherichia coli* : **A.** Lyophized vial. **B.** Growth on nutrient agar slant. **C.** Characteistic colonies on eosin methylene blue agar plate. **D.** Scanning electron micrographs at 15.00 KX magnifications.

enzyme, tryptophanase to break down the tryptophan, which makes by-products, of which, indole is the one. When Kovac's reagent (hydrochloric acid, p-dimethylaminobenzaldehyde and amyl alcohol) is added to a broth with indole in it, a dark pink color develops. The indole test must be read by 48 h of incubation because the indole can be further degraded if prolonged incubation occurs. The acidic pH produced by *Escherichia coli* limits its growth.

Method

- Inoculate a tube containing 5 ml of peptone water medium with presumptive colony taken from the nutrient agar plates under sterile conditions.
- Incubate at 37°C for 24 hrs.
- After incubation add 1.0 ml of the Kovac's reagent.
- Read results immediately.

Fig. 18. Indol test: **A.** Media control tube. **B.** Tube shows the purple ring, hence the test is positive.

Methyl Red Test

Principle

This test helps to differentiate between the species of the family Enterobacteriaceae. The test determines the ability of an organism to produce and maintain stable acid and products from glucose fermentation. *E. coli* is one of the bacteria that produce acids, causing the pH to drop below 4.4. When the pH indicator methyl red is added to this acidic broth it will be cherry red.

Method

- Inoculate glucose phosphate broth (MR-VP broth) with loopful of culture from nutrient agar plates.
- Incubate at 37°C for 24 hrs.
- Add about five drops of methyl red indicator solution.
- Record the results.

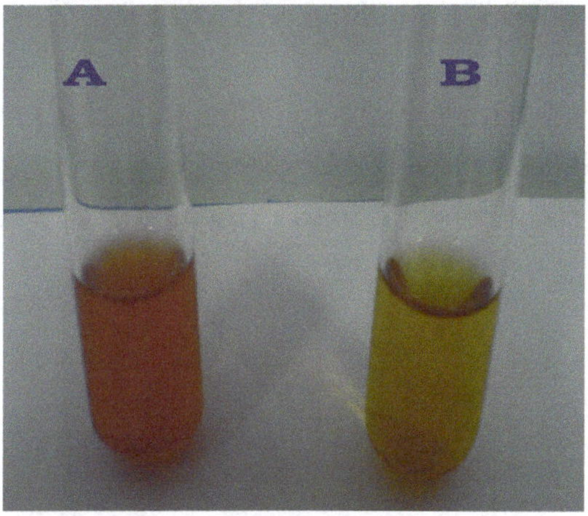

Fig. 19. Methyl Red Test: **A.** here red colour in test tube is indicative of positive test. **B.** Media control test tube.

Fermentation of Sugar
Principle

Most of bacteria will ferment a variety of sugars to form one or more acid end products.

Method

- Inoculate sugar medium with a loopful of culture from nutrient agar plate.
- Incubate it at 37°C for 24 hr.

Voges- Proskauer's test

This is a test for the production of acetylmethyl carbinol from glucose. To the inoculated medium after incubation, alkali is added, in the presence of which any acetylmethyl carbinol present becomes oxidized to diacetyl. The diacetyl will combine with creatine to give a red colour.

Method

- Inoculate glucose phosphate broth (MR-VP broth) with loopful of culture taken from nutrient agar plates
- Incubate broth for 24 h at 37°C.
- Transfer loopful of the culture from the inoculated broth into a clean test tube.
- Add 0.5 ml (8- 10 drops) of the α-naphthol solution and 0.5 ml of the 40% KOH solution containing 0.3% creatine.
- Shake thoroughly and allow to stand for 5 to 30 minutes.

Salmonella typhi

Kingdom	Bacteria
Phylum	Proteobacteria
Class	Gamma Proteobacteria
Order	Enterobacteriales
Family	Enterobacteriaceae
Genus	*Salmonella*
Species	*Salmonella typhi*

General Characteristics : The strains of *S. typhi* were first isolated in 1880 by Karl J. Erberth. This is a Gram-negative enteric bacillus belongs to the family Enterobacteriaceae. It is a motile, facultative anaerobe that is susceptible to various antibiotics and causes typhoid fever, paratyphoid fever, and foodborne illness (Giannella, 1996). Currently, 107 strains of this organism have been isolated; many containing varying metabolic characteristics, levels of virulence and multi-drug resistance genes that complicate treatment. *Salmonella* species are motile and produce hydrogen sulfide and has pink red colonies on Brilliant Green agar plate and black with metallic sheen on Bismuth sulphide agar plate.

The encounter of humans to *S. typhi* is made via faecal-oral route from infected individuals to healthy ones. Poor hygiene of patients

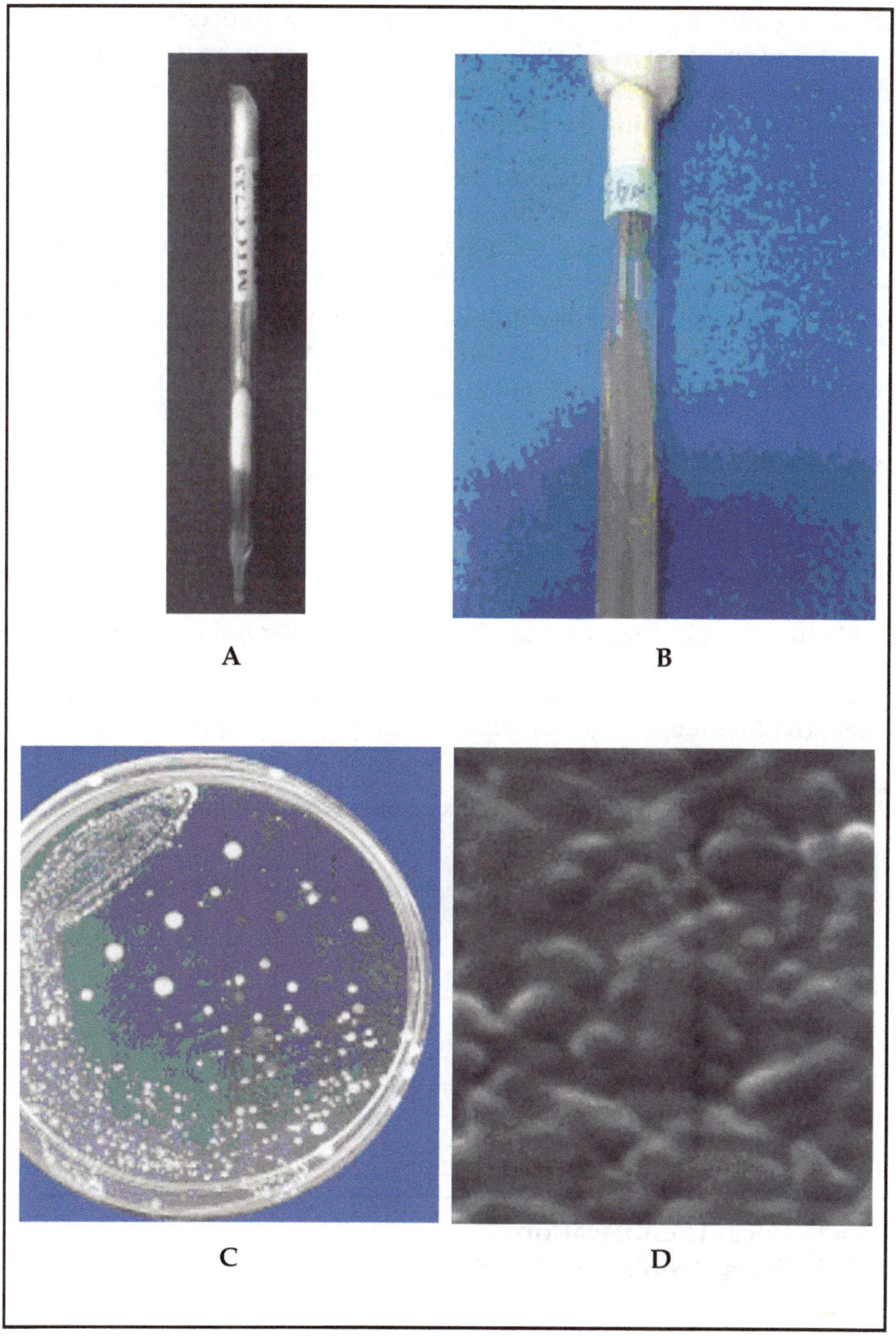

Fig. 20. Different features of *Salmonella typhi* : **A.** Lyophized vial. **B.** Growth on nutrient agar slant. **C.** Characterstics colonies on nutrient agar plate. **D.** Scanning electron micrographs at 12.00 KX magnification.

shedding the organism can lead to secondary infection, as well as consumption of shellfish from polluted bodies of water. The most common source of infection, however, is drinking water tainted by urine and faeces of infected individuals.

Infection of S. *typhi* leads to the development of typhoid, or enteric fever. This disease is characterized by the sudden onset of a sustained and systemic fever, severe headache, nausea, and loss of appetite. Other symptoms include constipation or diarrhoea, enlargement of the spleen, possible development of meningitis, and/or general malaise. Untreated typhoid fever cases result in mortality rates ranging from 12-30% while treated cases allow for 99% survival.

Macroscopic Examination

Characteristics colonies grow on specific type of media which are as follows

Table 10 : Morphological Characteristics of *Salmonella typhi* on Selective Agar Media

Selective Medium	Characteristic Colony Morphology	Gram Stain
Brilliant Green agar (BGA)	Reddish pink colonies	Gram-negative rods
Bismith sulphide agar (BSA)	Black with metallic sheen	Gram-negative rods
Xylose- Lysine Deoxycholate agar (XLD)	Red with black centre	Gram-negative rods
Deoxycholate Citrate agar	Colorless glassy opaque colonies	Gram-negative rods
MacConkey Agar	Colorless colonies	Gram-negative rods

Biochemical Identification

Triple Sugar Iron Agar Test (TSI)

Principle

This test is carried out to determine the ability of an organism to attack a specific carbohydrate incorporated into a basal growth medium by fermentation of glucose, lactose and sucrose present in it along with the determination of hydrogen sulphide (H_2S) production. Basically a pH indicator will change the colour of the media in response to fermentation. Colour change of the media in response to fermentation indicates which sugar is fermented. The presence of a black colour indicates that H_2S was produced. In this medium, H_2S reacts with the ferrous sulphate and convert it into ferrous sulfide that is black in colour.

Procedure

- With a sterile inoculating wire, inoculate suspect colonies into the blunt slant tube of Triple Sugar-Iron Agar Medium (By first streaking the surface of the slant and stabbing the wire beneath the surface).
- Incubate the biochemical media for 18-72 h at 37°C.

Observation

Butt Colour	Slant Colour	Result
Yellow	Pink slant and blackening due to H_2S	*Salmonella* present (Glucose only fermented)

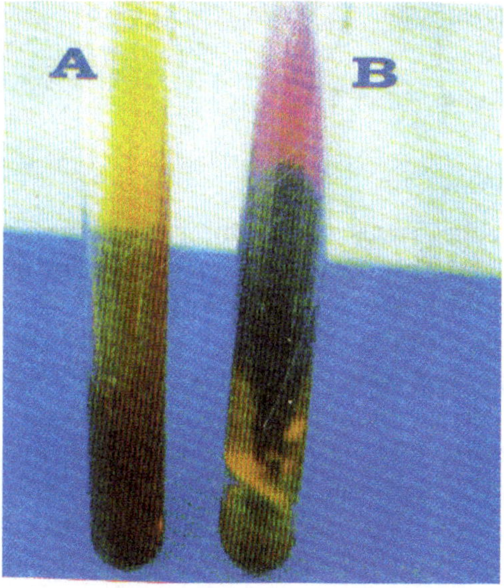

Fig. 21. *Triple Sugar Iron Agar Test:* **A.** Media control test tube. **B.** Tube shows the yellow butt, pink slant and blackening due to H_2S production (positive test)

Urease Test

Principle

This test is done to determine the ability of bacteria to hydrolyze urea to make ammonia using the enzyme urease. With phenol red as an indicator, the presence of urease is seen by the appearance of the red colour. This test helps to separate species of the genus Proteus from the urease-negative *Salmonella* and *Shigella* species.

Method

- Inoculate the slope of urease test medium (Urea agar base) with the colonies taken from nutrient agar plates.
- Incubate the inoculated tube for 18 to 24 hr at 37°C.

Observation

Colour of slant	Result
Red	Urease negative (no ammonia is produced), *S. typhi* present

Klebsiella pneumoniae

Kingdom	Bacteria
Phylum	Proteobacteria
Class	Gamma Proteobacteria
Order	Enterobacteriales
Family	Enterobacteriaceae
Genus	*Klebsiella*
Species	*K. pneumoniae*

General Characteristics : *Klebsiella* was named after the German bacteriologist Edwin Klebs (1834–1913). *K. pneumoniae* is a Gram-negative, non-motile, encapsulated, lactose fermenting, facultative anaerobic, rod shaped bacterium found in the normal flora of the mouth, skin, and intestines (Ryan and Ray, 2004). It is clinically the most important member of the *Klebsiella* genus of Enterobacteriaceae; it is closely related to *K. oxytoca* from which it is distinguished by being indole-negative and by its ability to grow on both melezitose and 3-hydroxybutyrate. It naturally occurs in the soil and about 30% of strains can fix nitrogen in anaerobic condition (Postgate, 1998). As a free-living *diazotroph*, its nitrogen fixation system has been much studied. New antibiotic resistant strains of *K. pneumoniae* are appearing, and it is increasingly found as a nosocomial infection. *K. pneumoniae* can cause bacterial pneumonia, typically due to aspiration by alcoholics, though it is more commonly implicated in hospital-acquired urinary tract and wound infections, particularly in immunocompromised individuals. Patients with *K. pneumoniae* tend to cough up characteristic sputum that is said to resemble "red-current jelly". *Klebsiella* ranks second to *E. coli* for urinary tract infections in older persons. It is also an opportunistic pathogen for patients with chronic pulmonary disease, enteric pathogenicity, nasal mucosa atrophy, and rhinoscleroma. Faeces are the most significant source of patient infection, followed by contact with contaminated instruments (Podschun and Ullman, 1998)

Macroscopic Examination

Characteristic colonies grow on specific type of media which are as follows

Table 11 : Morphological Characteristics of *Klebsiella pneumoniae* on Selective Agar Media

Selective Medium	Characteristic Colony Morphology	Gram Stain
MacConkey Agar Medium (MCA)	Pink colonies	Gram-Negative rods
Eosin Methylene Blue (EMB)	Pink mucoid	Gram-Negative rods

Citrate Utilization Test

Principle

To determine if an organism is capable of utilizing citrate as sole carbon source for metabolism with resulting alkalinity.

Method

- Streak a Simmon's Citrate agar slant with the organism and incubate at 37°C for 48 hours
- Examine for growth (+). Growth on the medium is accompianied by a rise in pH to change the medium from its initial green colour to deep blue.

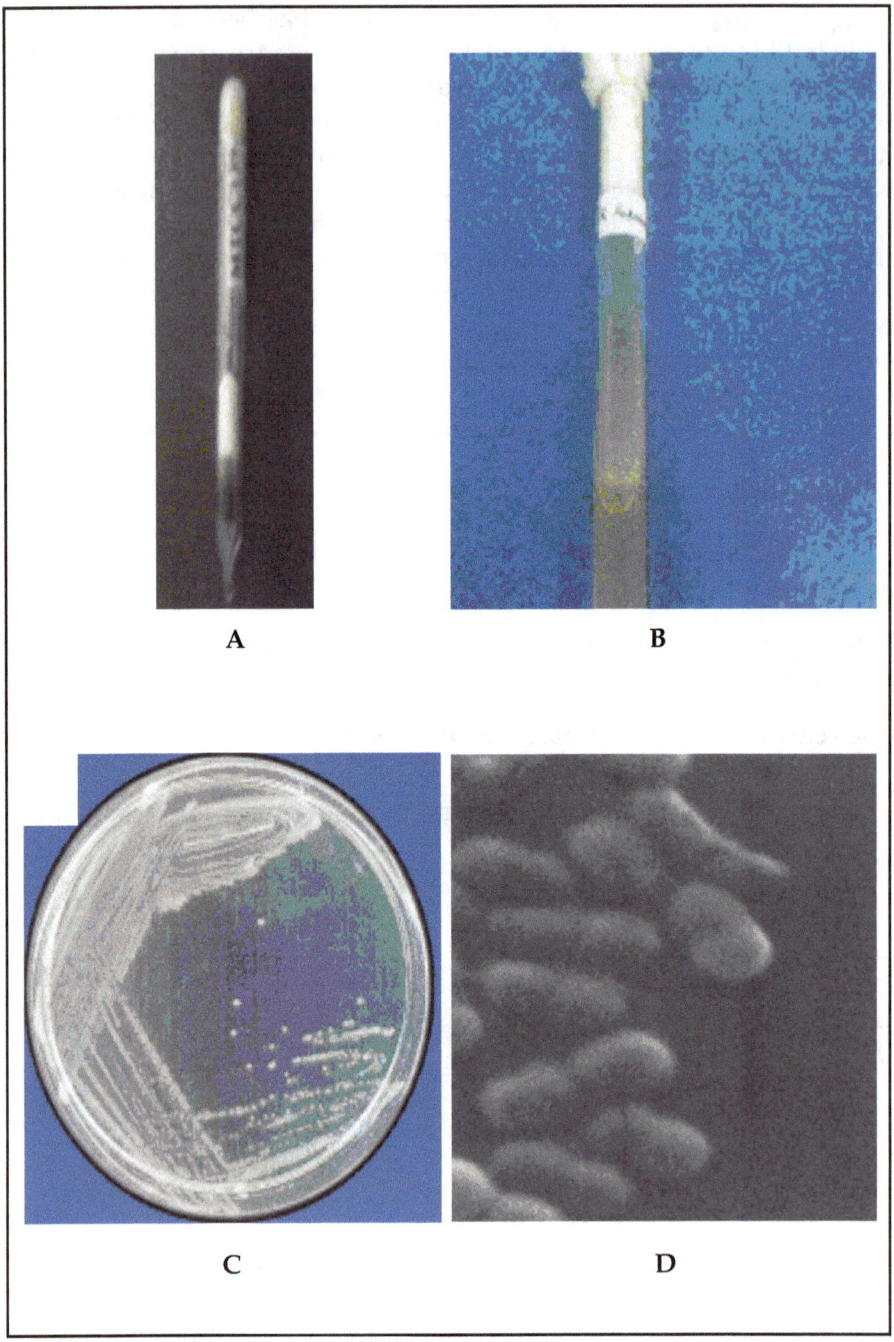

Fig. 22. Different features of *Klebsiella pneumoniae* : **A.** Lyophized vial. **B.** Growth on nutrient agar slant. **C.** Characteistic colonies on nutrient agar plate. **D.** Scanning electron micrographs at 12.00 KX magnifications.

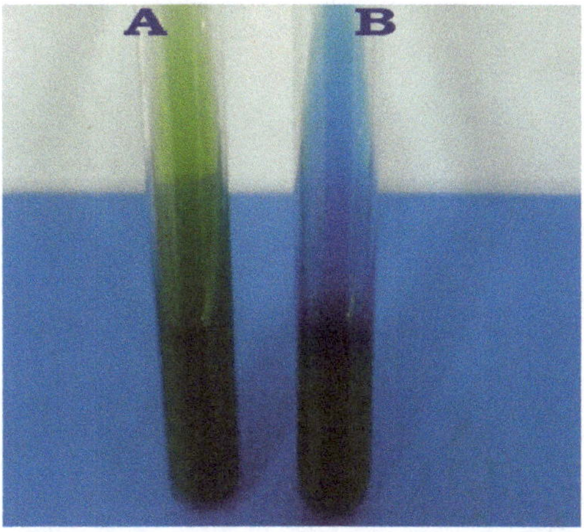

Fig. 23. *Citrate Utilization Test:* **A.** media control test tube. **B.** Tube shows the colour change from green to deep blue.

Table 12 : Results of Biochemical Test

S.No	Test	Results		
		E. coli	*S. typhi*	*K. pneumoniae*
1	Gram-staining	Gram-Negative	Gram-Negative	Gram-Negative
2	Indol	Positive	Negative	Negative
3	Methyl red	Positive	Positive	Positive
4	Voges -Proskauer	Negative	Negative	Positive
5	Citrate Utilization Test	Negative	Negative	Positive
6	Catalase	Positive	Positive	Positive
7	Oxidase	Negative	Negative	Negative
8	Glucose	Positive	Positive	Positive
9	Adonitol	Negative	Negative	Positive
10	Arabinose	Positive	Positive	Positive
11	Lactose	Positive	Negative	Positive
12	Sorbitol	Positive	Positive	Positive
13	Mannitol	Positive	Positive	Positive
14	Rhamnose	Positive	Negative	Positive
15	Sucrose	Positive	Negative	Positive

❑❑❑

Chapter 4

MATERIAL AND METHODS

Culture Media and Reagents

The media used throughout the study were procured from Hi-Media Laboratories, Mumbai, India and the chemicals used were procured locally. Composition of each medium is as follows:

1.0. Baird Parker agar

Ingredients	**g/l**
Casein enzymic hydrolysate	10.00
Beef extract	5.00
Yeast extract	1.00
Glycine	12.00
Sodium pyruvate	10.00
Lithium chloride	5.00
Agar	20.00
pH	7.0

2.0. Bismuth Sulphite Agar

Ingredients	g/l
Peptic digest of animal tissue	10.00
Beef extract	5.00
Dextrose	5.0
Disodium phosphate	4.0
Ferrous sulphate	0.30
Bismuth sulphite indicator	8.00
Brilliant Green	0.025
Agar	20.00
pH	7.7

3.0. Brilliant Green agar

Ingredients	g/l
Proteose peptone	10.00
Yeast extract	3.00
Lactose	10.0
Sucrose	10.0
Sodium chloride	5.00
Phenol Red	0.08
Brilliant Green	0.0125
Agar	20.00
pH	6.9

4.0. Cetrimide agar

Ingredients	g/l
Pancreatic digest of gelatin	20.00
Magnesium chloride	1.40
Potassium sulphate	10.00
Cetrimide	0.30
Agar	15.00
pH	7.2

5.0. Deoxycholate Citrate Agar

Ingredients	g/l
Heart infusion solids	10.00
Proteose peptone	10.00
Lactose	10.00
Sodium citrate	20.00
Ferric ammonium citrate	2.00
Sodium deoxycholate	5.00
Neutrel Red	0.02
Agar	15.00
pH	7.5

6.0. Hugh Leifson Medium

Ingredients	g/l
Peptic digest of animal tissue	2.00
Sodium chloride	5.00
Dipotassium phosphate	0.30
Glucose	10.00
Bromo thymol blue	0.05
Agar	2.00
pH	6.8

7.0. Eosine Methylene Blue Agar Medium

Ingredients	g/l
Peptone	10.00
Lactose	5.00
Sucrose	5.00
Dipotassium phosphate	2.0

Eosin Y	0.4
Methylene blue	0.065
Agar	15.00
pH	7.2

8.0. MacConkey agar

Ingredients	g/l
Peptic digest of animal tissue	17.00
Proteose peptone	3.00
Lactose	10.00
Bile salts	1.5
Sodium chloride	5.0
Neutral red	0.003
Agar	15
pH	7.1

9.0. Mannitol Salt agar

Ingredients	g/l
Proteose peptone	10.00
Beef extract	1.00
Sodium chloride	75.00
D-Mannitol	10.00
Phenol red	0.025
Agar	15.00
pH	7.4

10. Muller Hinton Agar

Ingredients	g/l
Beef infusion	3.00

Casein acid hydrolysate	17.5
Starch	1.50
Agar	17.00
pH	7.3

11. Muller Hinton Broth

Ingredients	**g/l**
Beef infusion	3.00
Infusion from Casein acid hydrolysate	17.5
Starch	1.50
pH	7.4

12. MR VP Medium

Ingredients	**g/l**
Buffered peptone	7.00
Dextrose	5.00
Dipotassium phosphate	5.00
pH	7.0

13. Nutrient Agar Media

Ingredients	**g/l**
Peptic digest of animal tissue	5.00
Beef extract	3.00
Yeast extract	1.50
Sodium chloride	5.00
Agar	15.00
pH	7.4

14. Nutrient Broth

Ingredients	g/l
Peptic digest of animal tissue	5.00
Beef extract	3.00
Yeast extract	1.50
Sodium chloride	5.00
pH	7.4

15. Peptone water

Ingredients	g/l
Peptic digest of animal tissue	10.00
Sodium chloride	5.0
pH	7.2

16. Simmons Citrate Agar

Ingredients	g/l
Magnesium sulphate	0.20
Ammonium dihydrogen phosphate	1.00
Dipotassium phosphate	1.00
Sodium citrate	2.00
Sodium chloride	5.00
Bromo thymol blue	0.08
Agar	15.00
pH	6.8

17. Triple Sugar Iron agar

Ingredients	g/l
Peptic digest of animal tissue	10.00
Casein enzymic hydrolysate	10.00

Yeast extract	3.00
Beef extract	3.00
Lactose	10.00
Sucrose	10.00
Dextrose	1.00
Sodium chloride	5.00
Ferrous sulphate	0.20
Sodium thiosulphate	0.30
Phenol red	0.024
Agar	12.00
pH	7.4

18. Urea Agar Base

Ingredients	**g/l**
Peptic digest of animal tissue	1.00
Dextrose	1.00
Sodium chloride	5.00
Disodium phosphate	1.20
Monopotassium phosphate	0.80
Phenol red	0.012
Agar	15.00
pH	6.8

19. Vogel-Johnson agar

Ingredients	**g/l**
Casein enzymic hydrolysate	10.00
Yeast extract	5.00
Mannitol	10.00

Dipotasium phosphate	5.00
Lithium chloride	5.00
Glycine	10.00
Phenol red	0.025
Agar	16.00
pH	7.2

20. Xylose-Lysine Deoxycholate Agar

Ingredients	**g/l**
Yeast extract	3.00
L-Lysine	5.0
Lactose	7.50
Sucrose	7.50.
Xylose	3.50
Sodium chloride	5.00
Sodium deoxycholate	2.50
Sodium thiosulphate	6.80
Ferric ammonium citrate	0.80
Phenol Red	0.08
Agar	15.00
pH	7.4

21. Kovac's Indol reagent

Ingredients	
p-dimethylamino benzaldehyde	5.0 g
Amyle alcohol	75.00 ml
Hydrochloric acid	25.00 ml

22. Alpha- Naphthylamine Solution

Ingredients	
Alpha- Naphthylamine	5.0 g
Acetic acid (5 N)	100 ml

23. Sulphanilic Acid (8%)

Ingredients	
Sulphanilic acid	8.0 g
Acetic acid (30 %)	1000ml

24. Solutions for Gram-staining

The stain consists of:

0.5% crystal violet in water

Solution A (Mix by addition of Crystal violet 2.0 g and 95% ethyl alcohol 20 ml.

Solution B (Mix by addition of Ammonium oxalate 0.8 g and 80% distilled water 80 ml.

Mix both the solution A and solution B.

Iodine solution (1g iodine and 2g potassium iodide in 300ml water)

Grams Decolourizer (0.5g iodine; 0.3g potassium iodide; 4.5ml methanol and 120ml acetone)

Safranin solution (2.5g safranin in 100ml of 95% ethyl alcohol)

25. Preparation of Cation adjusted MHB: (NCCLS, 1997)

Required cation content of Mg^{++} and Ca^{++} in Muller Hinton Broth is 10 mg/ml and 20 mg/ml, respectively.

Stock solution: 10 mg of Mg^{++} /ml

Dissolved 8.36 gm of $MgCl_2$ in 100 ml of distilled water

10 mg of Ca^{++} /ml

Dissolved 3.68 gm of $CaCl_2$ in 100 ml of distilled water

Stock solution was sterilized by membrane filtration and stored at 4°C.

Added 0.5 ml of stock solution of each in 1 litre of autoclaved MHB.

26. BG-11 Medium (Stainer *et al.*, 1971)

Component	g/l
Sodium nitrate	1.5
Dipotassium hydrogen phosphate	0.04
Magnesium sulphate	0.075
Calcium chloride	0.036
Citric acid	0.006
Ferric ammonium citrate	0.006
EDTA (di sodium magnecium salt)	0.001
Sodium carbonate	0.02
Trace metal mix	1 ml

Trace Metal Mix

Component	g/L
Boric acid	2.86
Manganese chloride	1.81
Zinc sulphate	0.222
Sodium molybdate	0.039
Copper sulphate	0.079
Cobalt nitrate	0.0494

Materials : Erlenmeyer flasks, centrifuge tube, glass rods, cover slips, glass slides, funnel, culture tube, measuring cylinder, petriplates, spirit lamp, forceps, pH strips, inoculation loop, test tube stand, pipettes, screw cap tube, well borer, reagent bottles, beaker, distlled water bottle cotton etc.

Equipments : Laminar air flow, oven, incubator, centrifuge, autoclave, weighing balance, microscope, refrigerator, shaker, water bath, sonicator, colony counter and Vernier callipers.

Sterilization of the Medium and Glassware

The medium was sterilized at 121°C and 15 lb/in^2 pressure for 15 minutes in an autoclave. The glassware like Petridishes, pipettes plugged with cotton at sucking end and cotton plugged culture tubes were wrapped in a paper and sterilized in a hot air oven at 160°C for 3 hrs.

Growth Curve for Cyanobacteria

Growth curves are generally prepared to study the progressive growth of cyanobacteria, and then they are examined to determine the early stationary phases of their life-cycle, which are believed to be the most appropriate periods for the production of secondary metabolites by an organism.

Light Scattering (Turbidity) Method

Light scattering method is the most general technique for measurement of the growth of pure cultures. This method is the time consuming as well as non-destructive for the growing cultures. The instrument used for this method is spectrophotometer, which generally measures the turbidity of the cultures as per their growth. OD of the experiment flask were taken just after adding the inoculum as the first reading and subsequently thereafter, with a gap period of two days, till the attainment of stationary phase by the culture. OD vs Number of days were plotted for the graphical representation of the life cycle of the organisms.

Extraction procedure : Collected cyanobacterial species were maintained in BG-11 growth medium (pH 7) under rotatory conditions and illuminated continuously at a light intensity of 3500 Lux. At the stationary phase of growth (23-25 days), the culture was harvested and spent medium was collected after filtering the biomass. The collected biomass was dried in a hot air oven at 60^{0}C. A known amount of dried biomass was then mixed properly in each solvent and then kept for few hours at room temperature. The mixture was further sonicated for 10 minutes and then centrifuged at 4000 rpm for 10 minutes. After centrifugation supernatant were collected in a preweighed test tube then concentrated under nitrogen until completely dry, again weighed and then resuspended in the appropriate solvent to make the solution of known concentration for the antibacterial assay.

Standard Antibiotic : Antibiotic (ciprofloxacin) used in the study was procured from Ranbaxy Research laboratories. On the basis of claim (500 mg/tablet), weighed accurately 0.150 g powder of ciprofloxacin into 100 ml of volumetric flask. Dissoloved and sonicated for five minutes and then made the volume to the mark with distilled water. Took 1 ml and diluted it to 100 ml. This dilution contained 10 μg/ml. Prepared double dilution to obtain 5 μg/ml. This solution was used as a standard solution throughout the study.

Antibacterial Assay

Prepration of Inoculum : The bacterial strains were inoculated on nutrient agar and incubated for 24 hrs at 30⁰C. After incubation, the cultures were suspended in saline solution (0.85% NaCl) and adjusted the number of test organisms approximately $1.0x10^7$ - $1.0x10^8$ cfu/ml by serial dilution up to 25% light transmittance at 530 nm by using spectrophotometer as per the guidelines given in Indian Pharmacopoeia (2007).

Agar well Diffusion Assay

In vitro antibacterial activity of different crude extracts of cyanobacterial species were evaluated using the agar well diffusion assay (Perez *et al.*, 1990). 100 μl of adjusted culture was mixed with 100 ml of Muller Hinton Agar (MHA), mixed well and poured 25 ml into each sterile Petridish (90 mm). This was then allowed to solidify and individual plates were marked for the organisms inoculated. After solidification, plates were punched to make the well of 6 mm diameter with the help of sterile borer. 100 μl of respective cyanobacterial extracts were pipetted into the well in assay plates. Plates were incubated overnight at 37⁰C. All the plates were observed for the zone of inhibition following the incubation, diameter of these zones were measured in millimeters. All tests were performed in sterile condition and repeated three times.

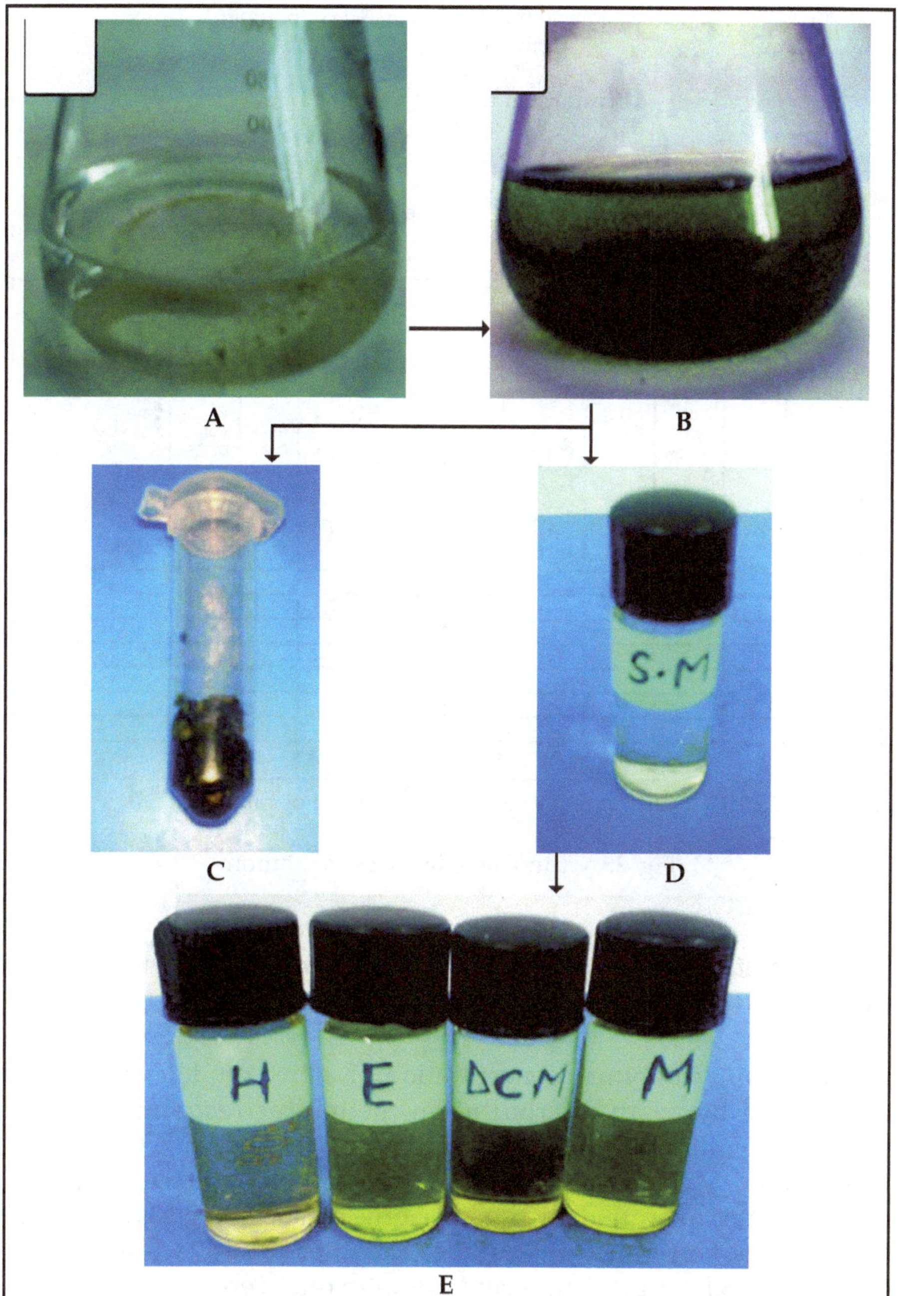

Fig. 24. Different stages of extraction : **A.** Zero day culture of cyanobacteria. **B.** Twenty five days old culture of cyanbacteria. **C.** Biomass after centrifugation **D.** Spent medium after centrifugation. **E.** Organic solvent extracts after sonication and centrifugation.

Done as recommended in the NCCLS document.
Stock of 10 mg/ml
↓
Diluted 1:5 to get 2000 μg/ml
↓
To get 1024 μg/ml (512 μl of 1:5 diluted stock + 488 μl of MHB)
↓
Added 200 μl of this in the second column

	1 200 μl MHB (-ve control)	2 200 μl Extract	3 50 μl MHB + 50 μl Extract from Column 2	4 75 μl MHB + 25 μl Extract from Column 2	**5 175 μl MHB + 25 μl Extract from Column 2**	6 50 μl MHB + 50 μl Extract from Column 5	7 75 μl MHB + 25 μl Extract from Column 5	**8 175 μl MHB + 25 μl Extract from Column 5**	9 50 μl MHB + 50 μl Extract from Column 8	10 75 μl MHB + 25 μl Extract from Column 8	**11 175 μl MHB + 25 μl Extract from Column 8**	12 100 μl MHB + 100μl of culture **(+ve control)**
A												
B												
C												
D												
E												
F												
G												
H												

Remove 100 μl from column no.11 before adding 100 μl of culture

Fig. 25. Cyanobacterial extracts dilution

Determination of Minimum Inhibitory Concentration (MIC) by Broth Dilution Method

Minimum inhibitory concentration of active crude extracts was determined by broth microdilution method as recommended by National Committee for Clinical Laboratory Standards (NCCLS, 1997). This method is called "microdilution" because it involves the use of small volumes of broth dispensed in sterile plastic microdilution plate. The test was performed in 96 wells microtiter plates, two fold serial dilutions of all extracts and standard antibiotic were made in Cation Adjusted Muller-Hinton Broth (CAMHB) ranging from 1 to 512 μg/ml. Each inoculum was prepared in the same medium at a density adjusted to a 25% transmittance turbidity standard (10^8 cfu/ml) and diluted to 1:100 for the assay. After inoculation, each well contained approximately 5 x 10^5 cfu/ml. The final volume in the wells was 200 µl. After 24 hrs of incubation at 37^{0}C, the calculated amount of nitrogen dried cyanobacterial material present in the most diluted extract that inhibit a visible growth was defined as MIC.

Chapter 5

RESULTS

Nine cyanobacterial strains *viz., Anabaena variabilis, A. fertilissima, Nostoc muscorum, N. punctiforme, N. linckia, N. commune, Spirulina platensis, Westiellopsis prolifica* and *Hapalosiphon sp.* were cultured in the laboratory. The growth curve for each culture was studied by growing them in Erlenmeyer flask using BG-11 growth medium. They were examined to determine the early stationary phases of their life-cycle, which are believed to be the most appropriate period for the production of secondary metabolites by an organism. After 23-25 days of incubation period, each culture was harvested and spent medium was collected by filtration of the biomass. Hexane, ethyl acetate, dichloromethane and methanol were used as the solvents for obtaining the extracts from biomass and spent medium of each culture which were then investigated against five pathogenic bacteria *viz., Staphylococcus aureus, Escherichia coli, Pseudomonas aeruginosa, Klebsiella pneumoniae* and *Salmonella typhi*. Antibacterial activity of different cyanobacterial sp. has been shown in various tables.

Table 13 : Growth curve readings of selected Cyanobacteria

No.of days	Optical Density(OD) at 650 nm								
	Anabaena variabilis	*Anabaena fertilissima*	*Nostoc muscorum*	*Nostoc punctiforme*	*Nostoc linckia*	*Nostoc commune*	*Spirulina platensis*	*Westiellopsis prolifica*	*Hapalosiphon sp.*
0	0.066	0.07	0.059	0.064	0.05	0.063	0.3	0.08	0.044
2	0.084	0.099	0.078	0.08	0.077	0.085	0.4	0.09	0.06
4	0.088	0.121	0.08	0.088	0.083	0.09	0.43	0.13	0.08
6	0.133	0.144	0.121	0.117	0.128	0.115	0.444	0.15	0.133
8	0.129	0.16	0.188	0.179	0.171	0.19	0.51	0.199	0.193
10	0.209	0.2	0.223	0.2	0.229	0.232	0.58	0.25	0.24
12	0.33	0.38	0.35	0.355	0.345	0.333	0.795	0.24	0.38
14	0.588	0.699	0.666	0.66	0.658	0.67	0.925	0.666	0.65
16	0.777	0.81	0.799	0.79	0.795	0.78	1.41	0.8	0.71
18	0.892	0.899	0.823	0.927	0.916	0.91	1.4	0.92	0.72
20	0.897	0.912	0.812	0.914	0.912	0.88	1.39	0.89	0.689
22	0.924	0.89	0.867	0.922	0.923	0.923	1.46	0.934	0.712
24	0.914	0.922	0.845	0.934	0.945	0.934	1.4	0.913	0.705
26	0.957	0.936	0.857	0.943	0.934	0.913	1.45	0.924	0.723

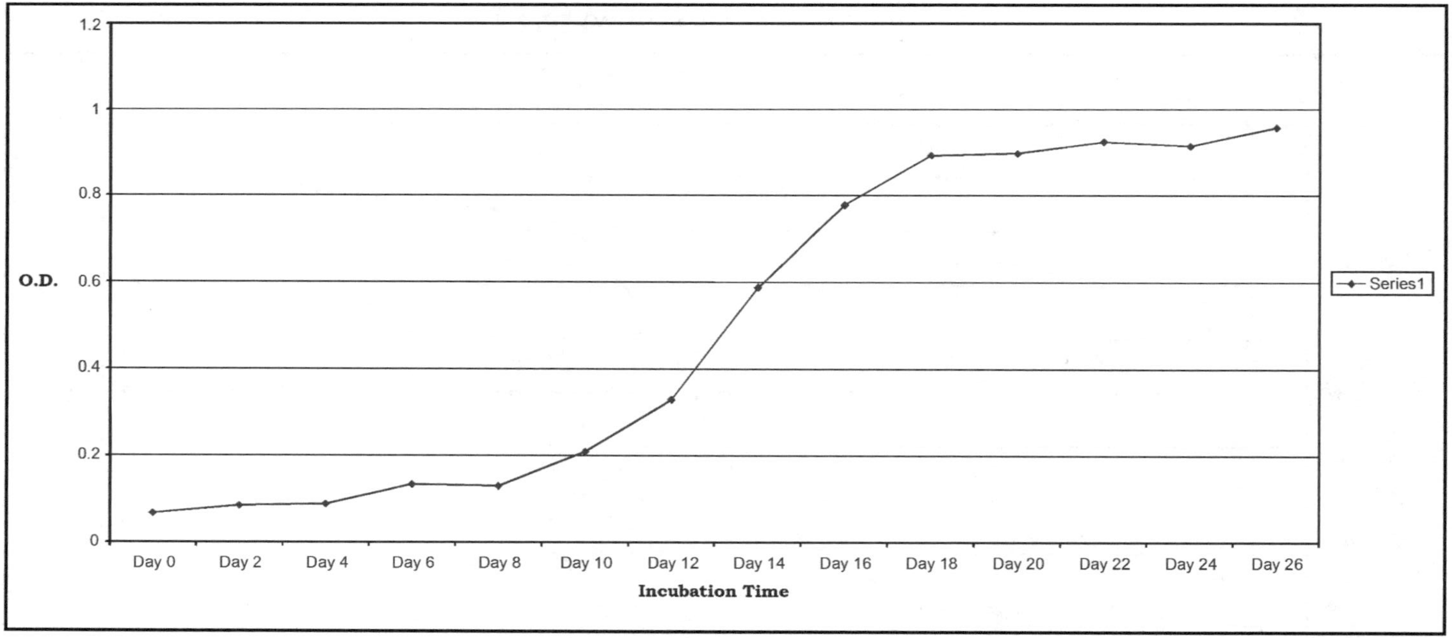

Fig. 26. Growth curve for *Anabaena variabilis*

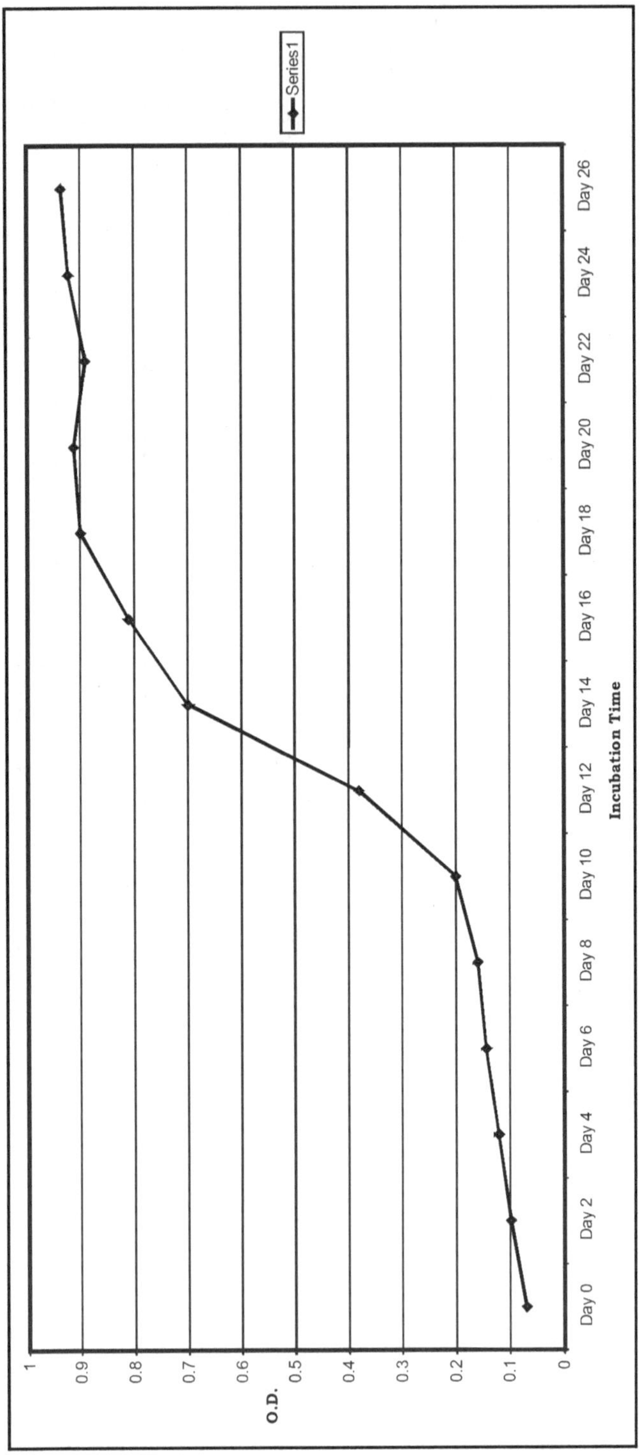

Fig. 27. Growth curve for *Anabaena fertilissima*

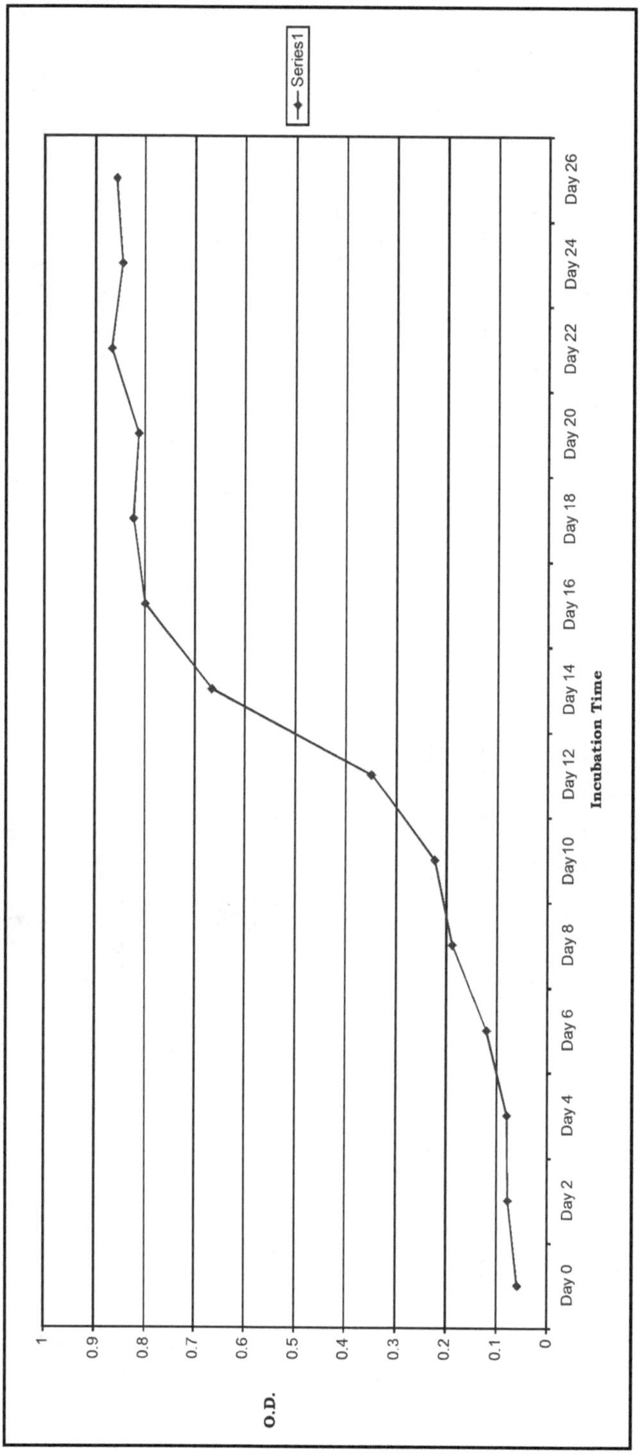

Fig. 28. Growth curve for *Nostoc muscorum*

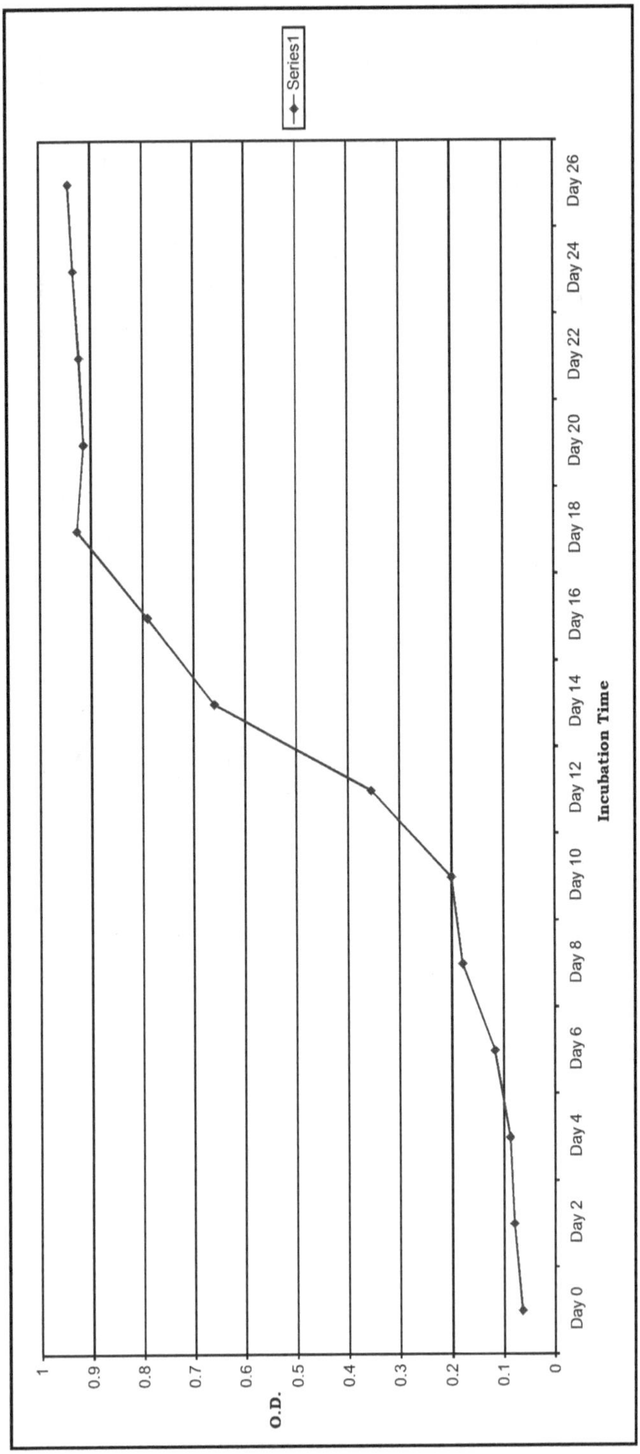

Fig. 29. Growth curve for *Nostoc punctiforme*

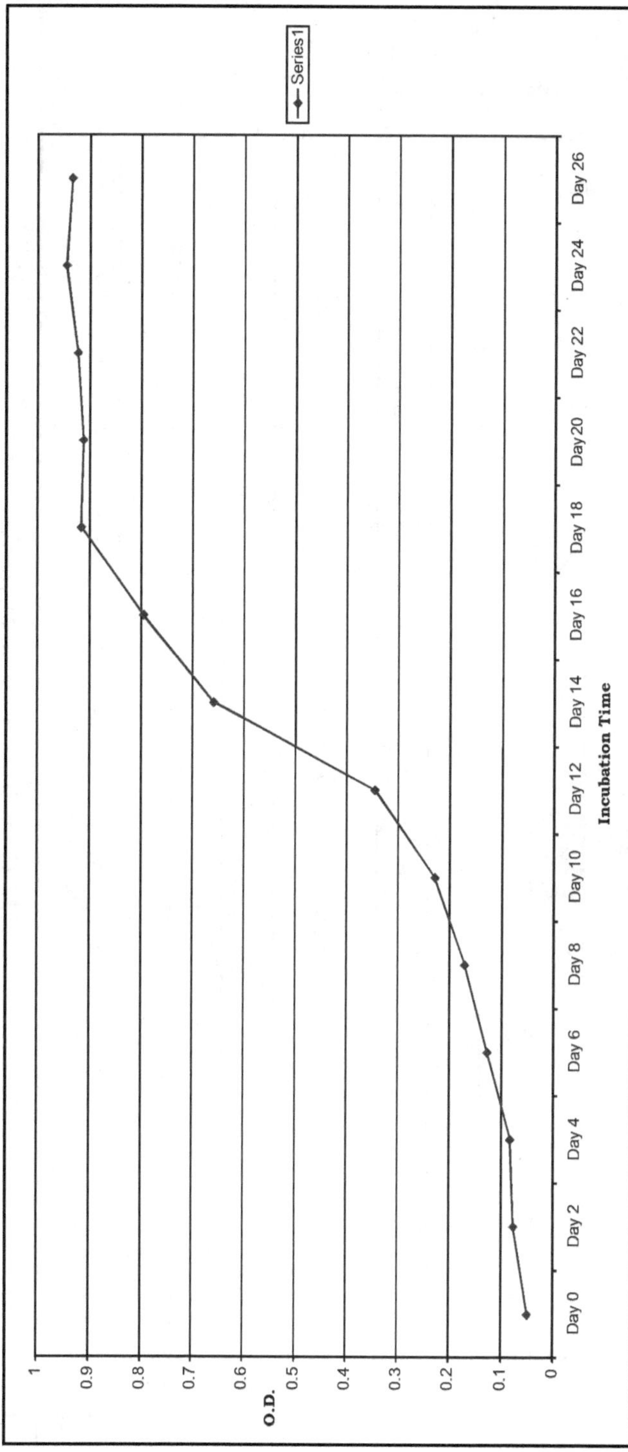

Fig. 30. Growth curve for *Nostoc linckia*

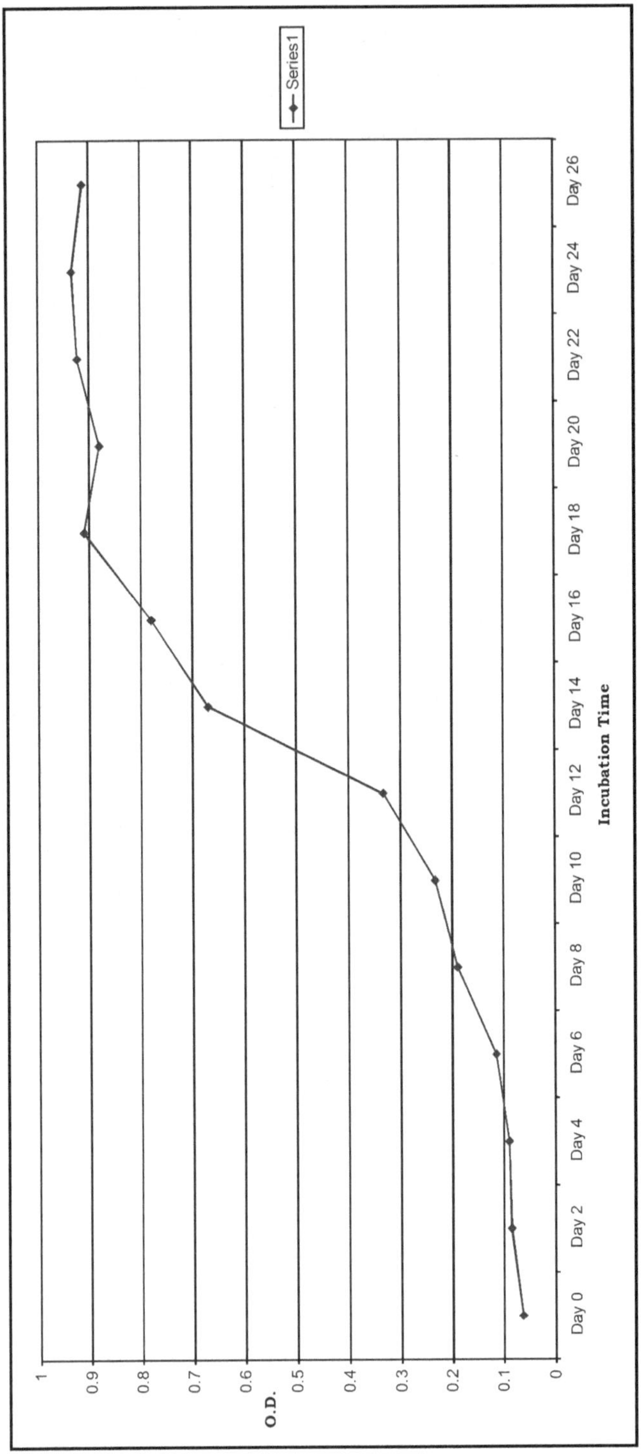

Fig. 31. Growth curve for *Nostoc commune*

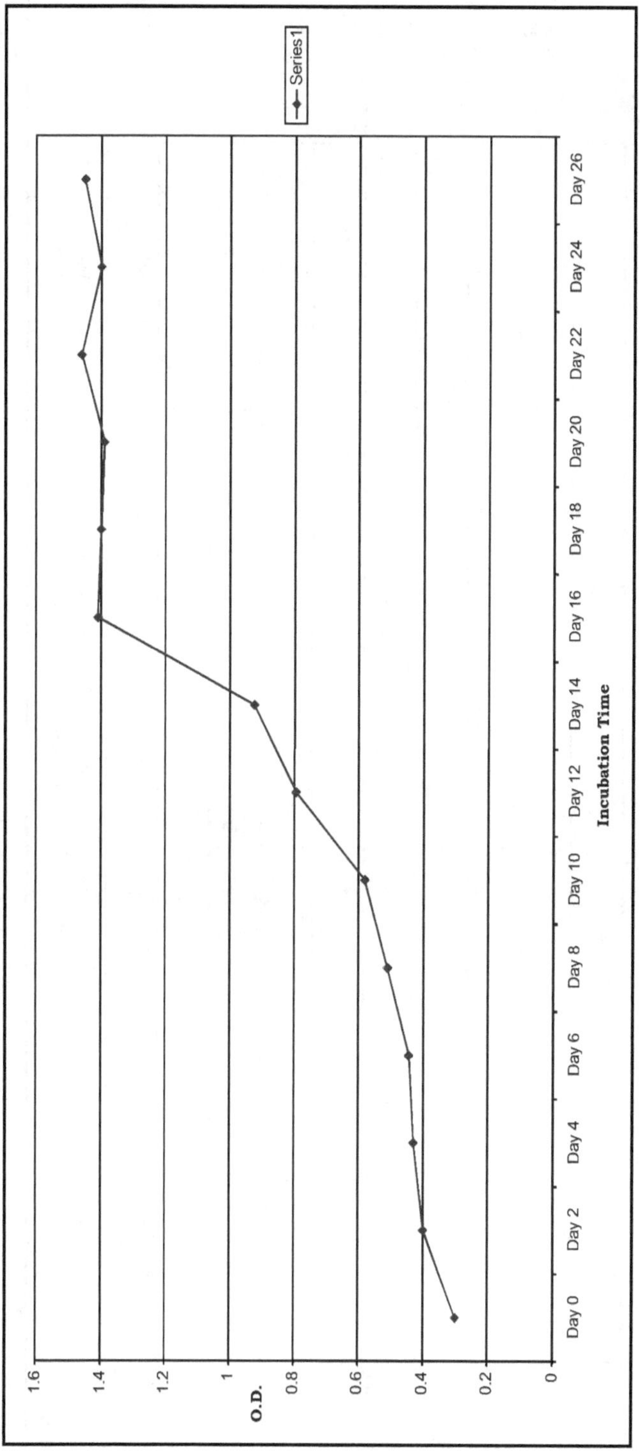

Fig. 32. Growth curve for *Spirulina platensis*

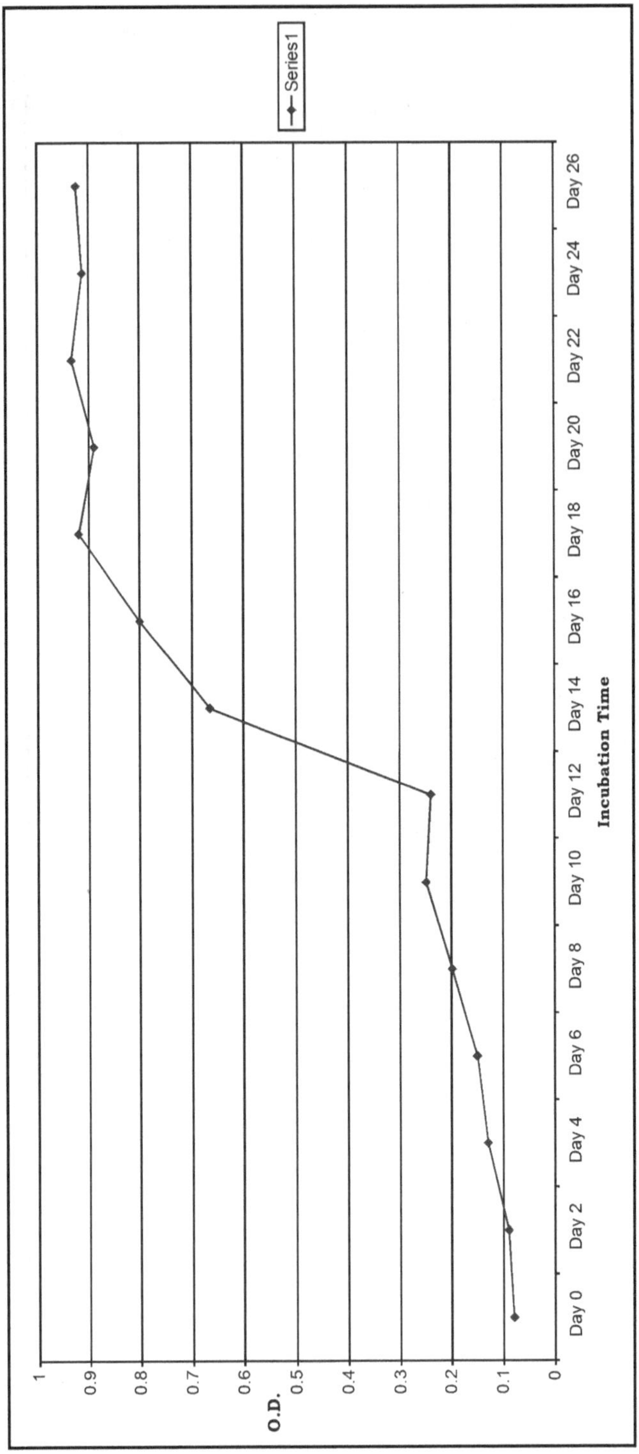

Fig. 33. Growth curve for *Westiellopsis prolifica*

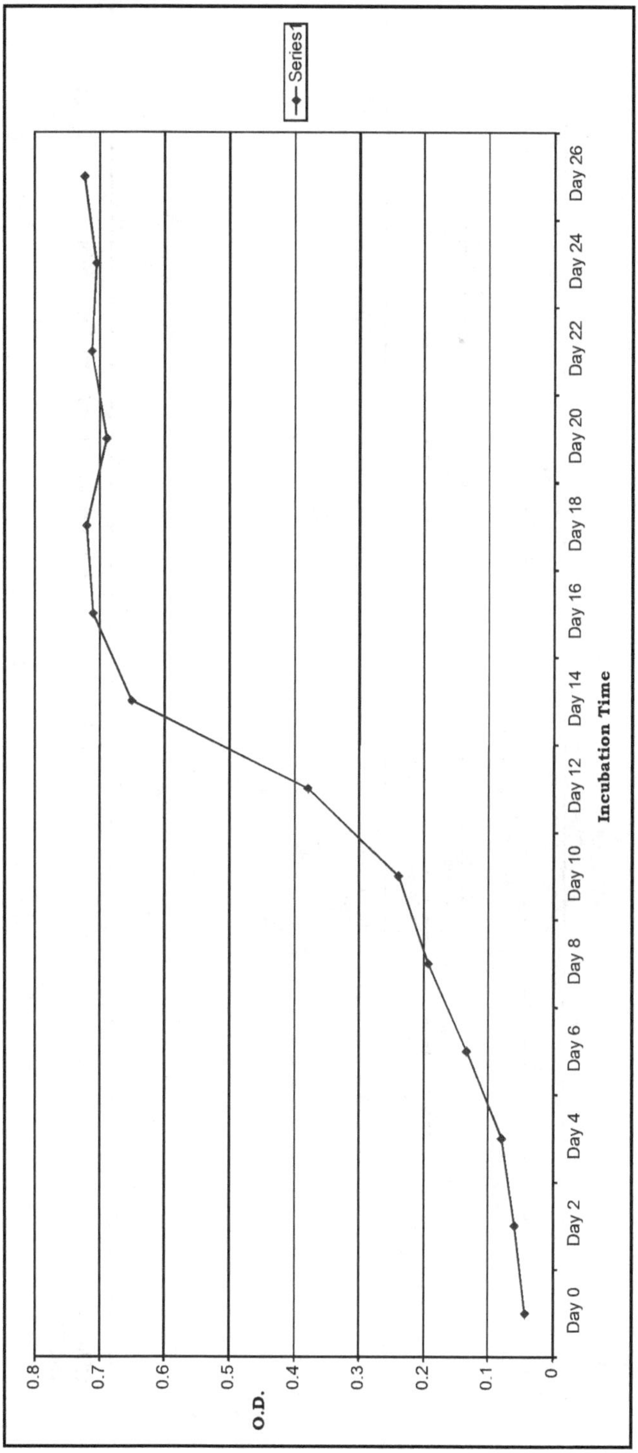

Fig. 34. Growth curve for *Hapalosiphon sp.*

Table 14. Antibacterial activity of different extracts of *Anabaena varibilis*

Test organisms	Effective zone of inhibition (mm)										
	Organic Extracts (10mg/ml)								Extracellular substances		
	Hexane		Ethyl acetate		Dichloromethane		Methanol				
	Extract	Control	Extract	Control	Extract	Control	Extract	Control	Spent medium	BG-11 Media Control	Ciprofloxacin (5 µg/ml)
Staphylococcus aureus	4.31±.35	0.0	6.42±.53	0.0	0.0	0.0	10.14±2.1	0.0	0.0	0.0	25.33±.15
Escherichia coli	0.0	0.0	5.41±.3	0.0	0.0	0.0	8.92±1.4	0.0	0.0	0.0	23.46±.4
Pseudomonas aeruginosa	0.0	0.0	0.0	0.0	0.0	0.0	4.48±1.2	0.0	0.0	0.0	21.35±.22
Klebsiella pneumoniae	0.0	0.0	0.0	0.0	0.0	0.0	0.0	0.0	0.0	0.0	17.35±.22
Salmonella typhi	0.0	0.0	0.0	0.0	0.0	0.0	8.35±1	0.0	0.0	0.0	28.44±.28

Results are the means of diameter values ± standard deviation. Effective zone of inhibition (total zone of inhibition, minus diameter of well)

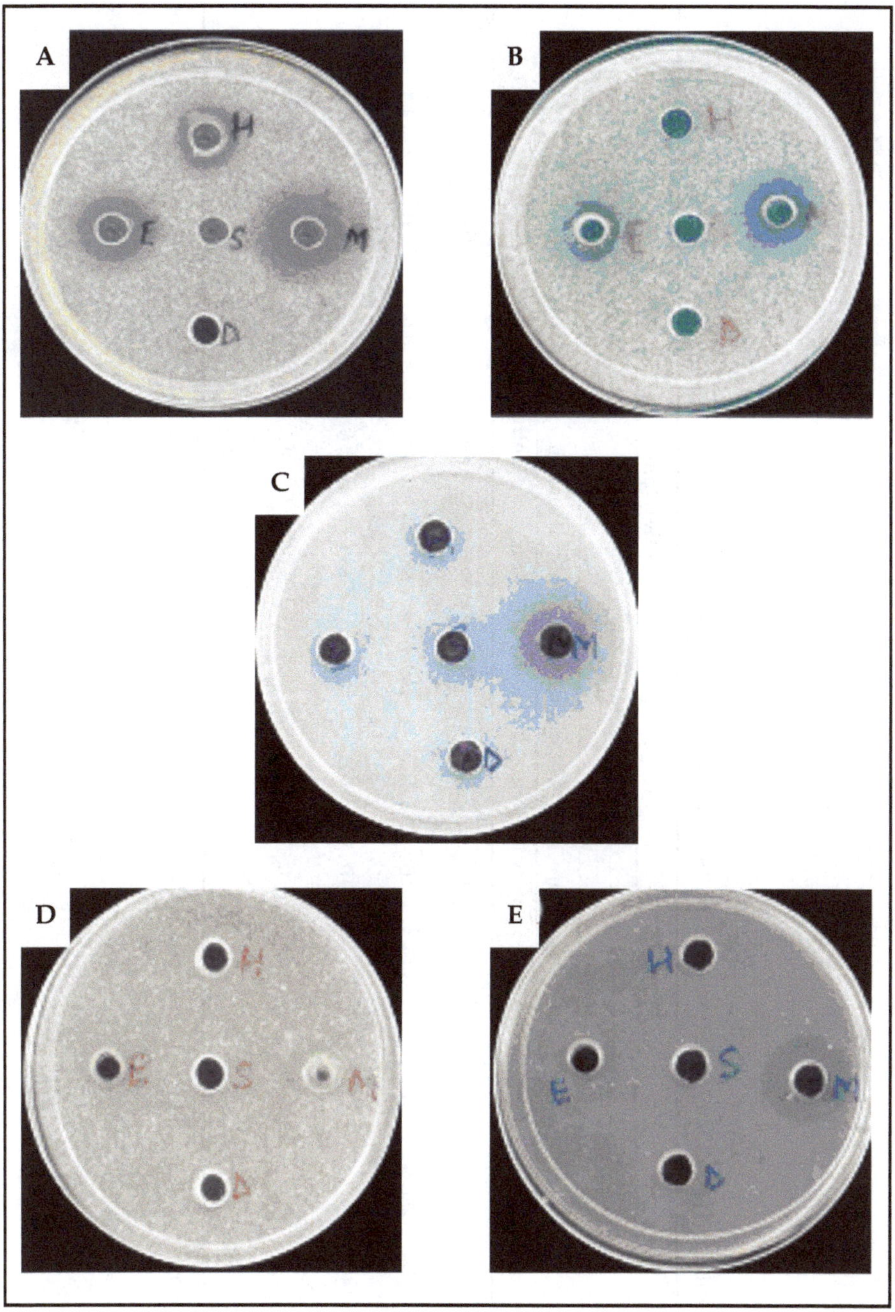

Fig. 35. Zone of inhibition exhibited by different extracts of *Anabaena variabilis*. **A.** *S. aureus*; **B.** *E. coli.*; **C.** *P aeruginos*; **D.** *K. pneumoniae*; **E.** *S. typhi*, Inside Petriplates : H, hexane extract ; E, ethyl acetate extract; D, dichloromethane extract; M, methanolic extract; S, spent medium

Table 15. Antibacterial activity of different extracts of *Anabaena fertilissima*

Test organisms	Effective zone of inhibition (mm)										
	Organic Extracts (10mg/ml)								Extracellular substances		
	Hexane		Ethyl acetate		Dichloromethane		Methanol				
	Extract	Control	Extract	Control	Extract	Control	Extract	Control	Spent medium	BG-11 Media Control	Ciprofloxacin (5 μg/ml)
Staphylococcus aureus	0.0	0.0	5.37±.78	0.0	0.0	0.0	8.56±1.1	0.0	0.0	0.0	25.33±.15
Escherichia coli	0.0	0.0	4.42±13	0.0	0.0	0.0	7.59±.63	0.0	0.0	0.0	23.46±.4
Pseudomonas aeruginosa	0.0	0.0	0.0	0.0	0.0	0.0	4.36±1.4	0.0	0.0	0.0	21.35±.22
Klebsiella pneumoniae	0.0	0.0	0.0	0.0	0.0	0.0	0.0	0.0	0.0	0.0	17.35±.22
Salmonella typhi	0.0	0.0	0.0	0.0	0.0	0.0	3.38±1.7	0.0	0.0	0.0	28.44±.28

Results are the means of diameter values ± standard deviation. Effective zone of inhibition (total zone of inhibition, minus diameter of well)

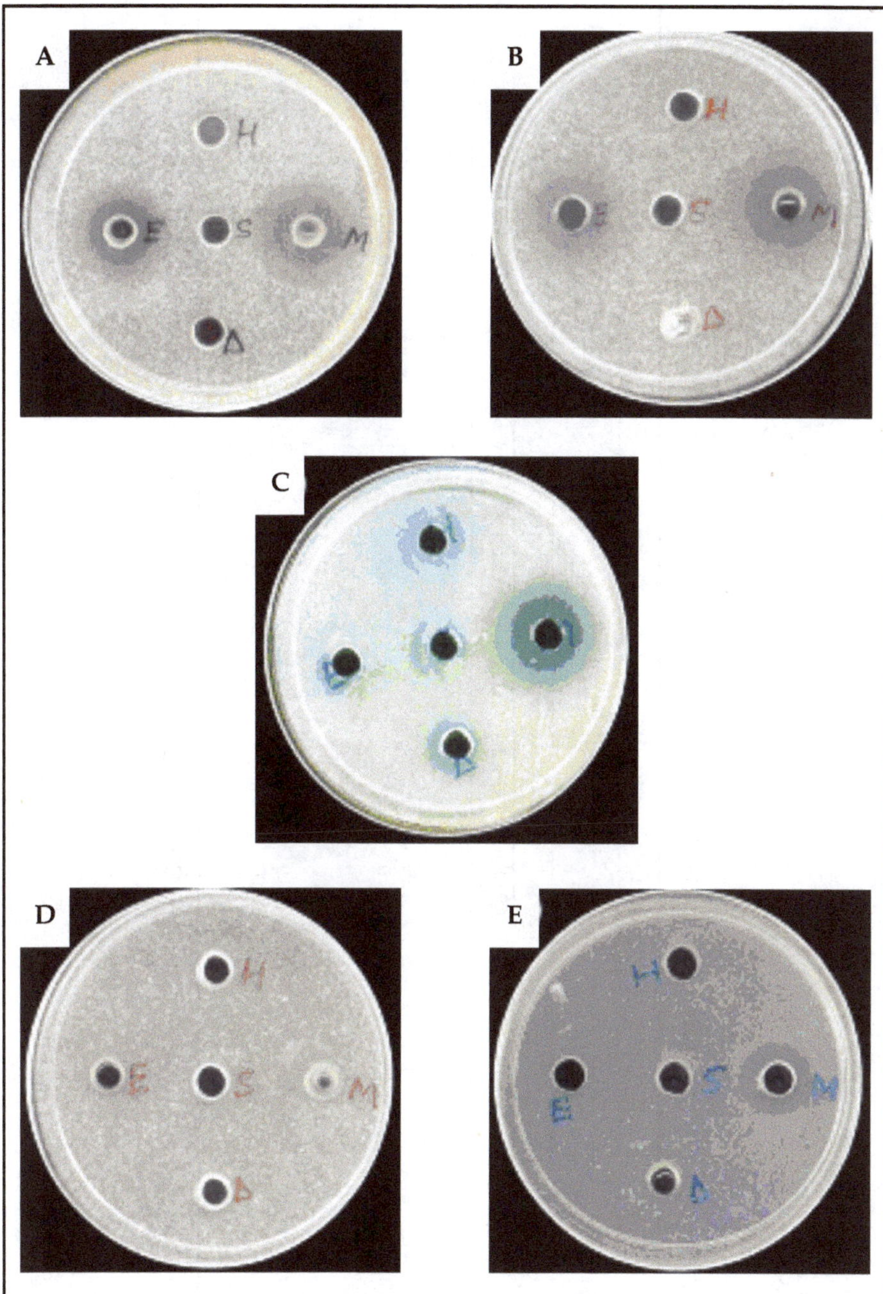

Fig. 36. Zone of inhibition exhibited by different extracts of *Anabaena fertilissima*. **A.** *S. aureus*; **B.** *E. coli*; **C.** *P. aeruginosa* ; **D.** *K. pneumoniae* ; **E.** *S. typhi*. **Inside Petriplates** : H, hexane extract ; E, ethyl acetate extract; D, dichloromethane extract; M, methanolic extract; S, spent medium

Table 16. Antibacterial activity of different extracts of *Nostoc muscorum.*

Test organisms	Effective zone of inhibition (mm)										
	Organic Extracts (10mg/ml)								Extracellular substances		
	Hexane		Ethyl acetate		Dichloromethane		Methanol				
	Extract	Control	Extract	Control	Extract	Control	Extract	Control	Spent medium	BG-11 Media Control	Ciprofloxacin (5 µg/ml)
Staphylococcus aureus	3.32±1.5	0.0	8.50±1.7	0.0	0.0	0.0	11.72±3	0.0	0.0	0.0	25.33±.15
Escherichia coli	0.0	0.0	7.33±.9	0.0	0.0	0.0	9.46±1.5	0.0	0.0	0.0	23.46±.4
Pseudomonas aeruginosa	0.0	0.0	0.0	0.0	0.0	0.0	8.45±1.4	0.0	0.0	0.0	21.35±.22
Klebsiella pneumoniae	0.0	0.0	0.0	0.0	0.0	0.0	0.0	0.0	0.0	0.0	17.35±.22
Salmonella typhi	0.0	0.0	0.0	0.0	0.0	0.0	5.09±.8.0	0.0	0.0	0.0	28.44±.28

Results are the means of diameter values ± standard deviation. Effective zone of inhibition (total zone of inhibition, minus diameter of well)

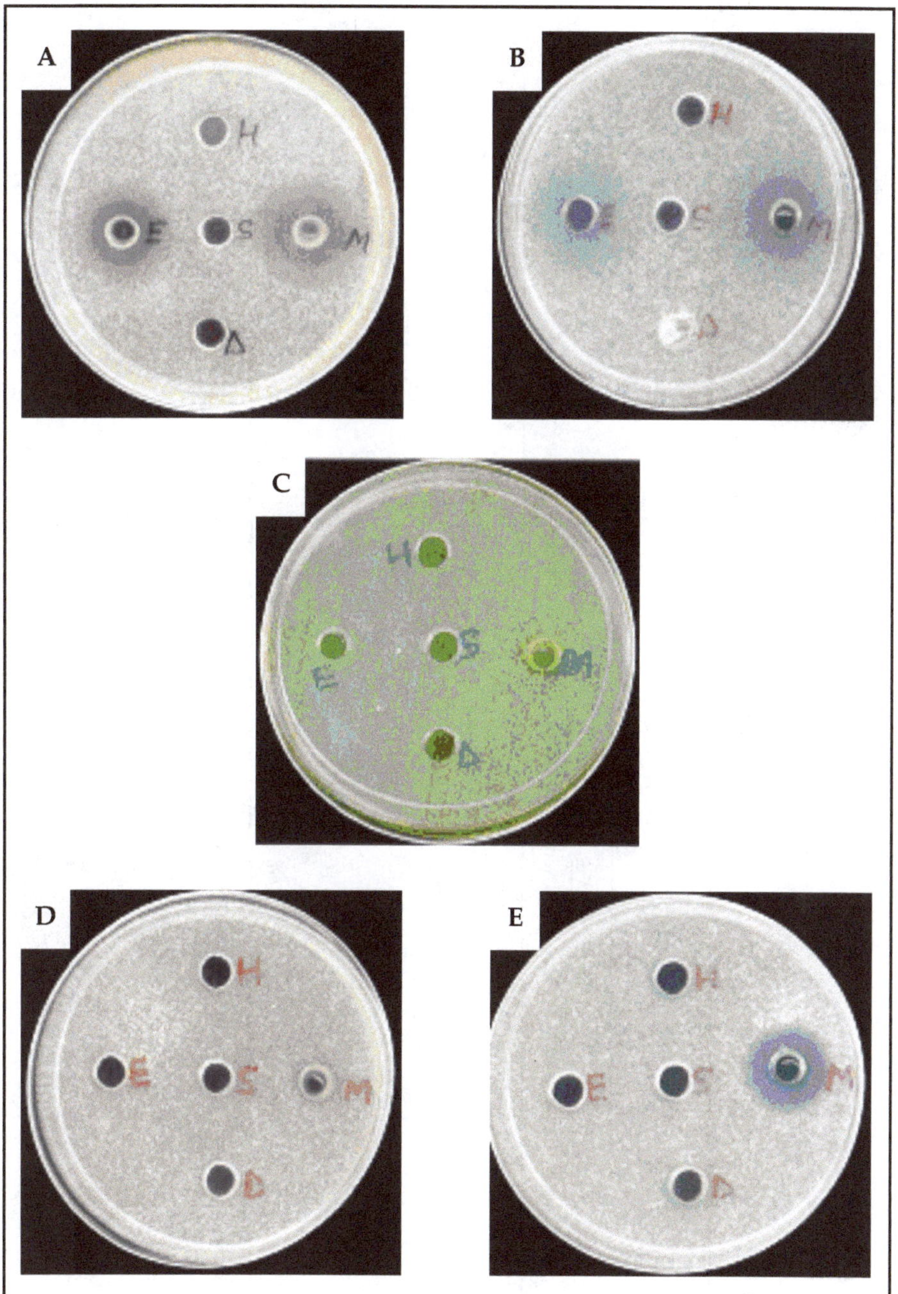

Fig. 37. Zone of inhibition exhibited by different extracts of *Nostoc muscorum* **A.** *S. aureus* ; **B.** *E. coli* ; **C.** *P. aeruginosa* ; **D.** *K. pneumoniae* ; **E.** *S. typhi*. **Inside Petriplate** : H, hexane extract ; E, ethyl acetate extract; D, dichloromethane extract; M, methanolic extract; S, spent medium.

Table 17. Antibacterial activity of different extracts of *Nostoc punctiforme*.

Test organisms	Effective zone of inhibition (mm)										
	Organic Extracts (10mg/ml)								Extracellular substances		
	Hexane		Ethyl acetate		Dichloromethane		Methanol				
	Extract	Control	Extract	Control	Extract	Control	Extract	Control	Spent medium	BG-11 Media Control	Ciprofloxacin (5 µg/ml)
Staphylococcus aureus	0.0	0.0	5.44±1	0.0	0.0	0.0	7.58±.7	0.0	0.0	0.0	25.33±.15
Escherichia coli	0.0	0.0	4.32±1.5	0.0	0.0	0.0	6.62±.5	0.0	0.0	0.0	23.46±.4
Pseudomonas aeruginosa	0.0	0.0	4.36±.5	0.0	0.0	0.0	4.48±1.2	0.0	0.0	0.0	21.35±.22
Klebsiella pneumoniae	0.0	0.0	0.0	0.0	0.0	0.0	0.0	0.0	0.0	0.0	17.35±.22
Salmonella typhi	0.0	0.0	0.0	0.0	0.0	0.0	4.37±1.4	0.0	0.0	0.0	28.44±.28

Results are the means of diameter values ± standard deviation. Effective zone of inhibition (total zone of inhibition, minus diameter of well)

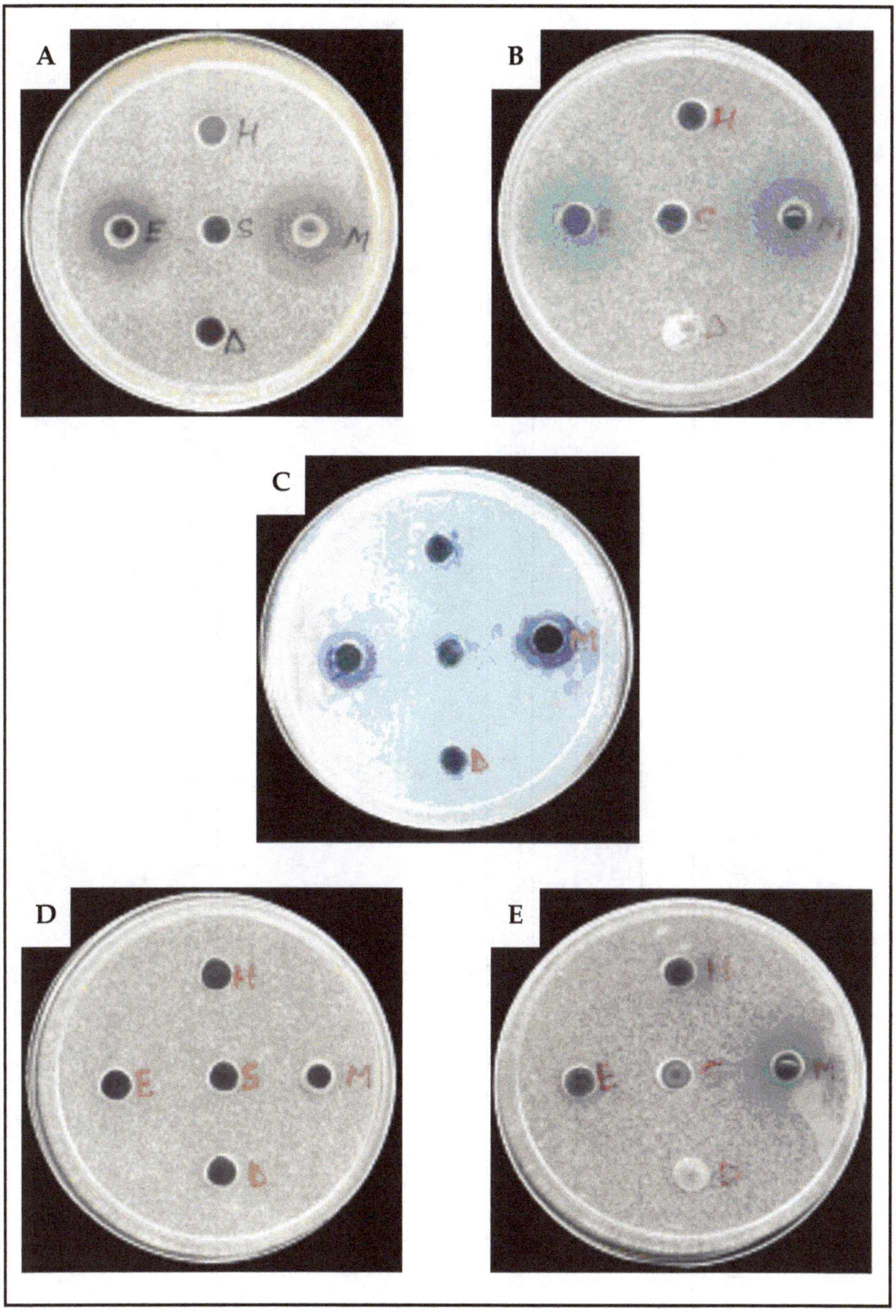

Fig. 38. Zone of inhibition exhibited by different extracts of *Nostoc punctiforme* **A.** *S. aureus;* **B.** *E. coli* ; **C.** *P. aeruginosa* ; **D.** *K. pneumoniae* ; **E.** *S. typhi.* **Inside Petriplate** : H, hexane extract ; E, ethyl acetate extract; D, dichloromethane extract; M, methanolic extract; S, spent medium

Table 18. Antibacterial activity of different extracts of *Nostoc linckia.*

Test organisms	Effective zone of inhibition (mm)										
	Organic Extracts (10mg/m l)								Extracellular substances		
	Hexane		Ethyl acetate		Dichloromethane		Methanol				
	Extract	Control	Extract	Control	Extract	Control	Extract	Control	Spent medium	BG-11 Media Control	Ciprofloxacin (5µg/ml)
Staphylococcus aureus	0.0	0.0	6.96±.6	0.0	0.0	0.0	8.59±1.1	0.0	0.0	0.0	25.33±.15
Escherichia coli	0.0	0.0	0.0	0.0	0.0	0.0	4.28±1.4	0.0	0.0	0.0	23.46±.4
Pseudomonas aeruginosa	0.0	0.0	0.0	0.0	0.0	0.0	0.0	0.0	0.0	0.0	21.35±.22
Klebsiella pneumoniae	0.0	0.0	0.0	0.0	0.0	0.0	0.0	0.0	0.0	0.0	17.35±.22
Salmonella typhi	0.0	0.0	0.0	0.0	0.0	0.0	0.0	0.0	0.0	0.0	28.44±.28

Results are the means of diameter values ± standard deviation. Effective zone of inhibition (total zone of inhibition, minus diameter of well)

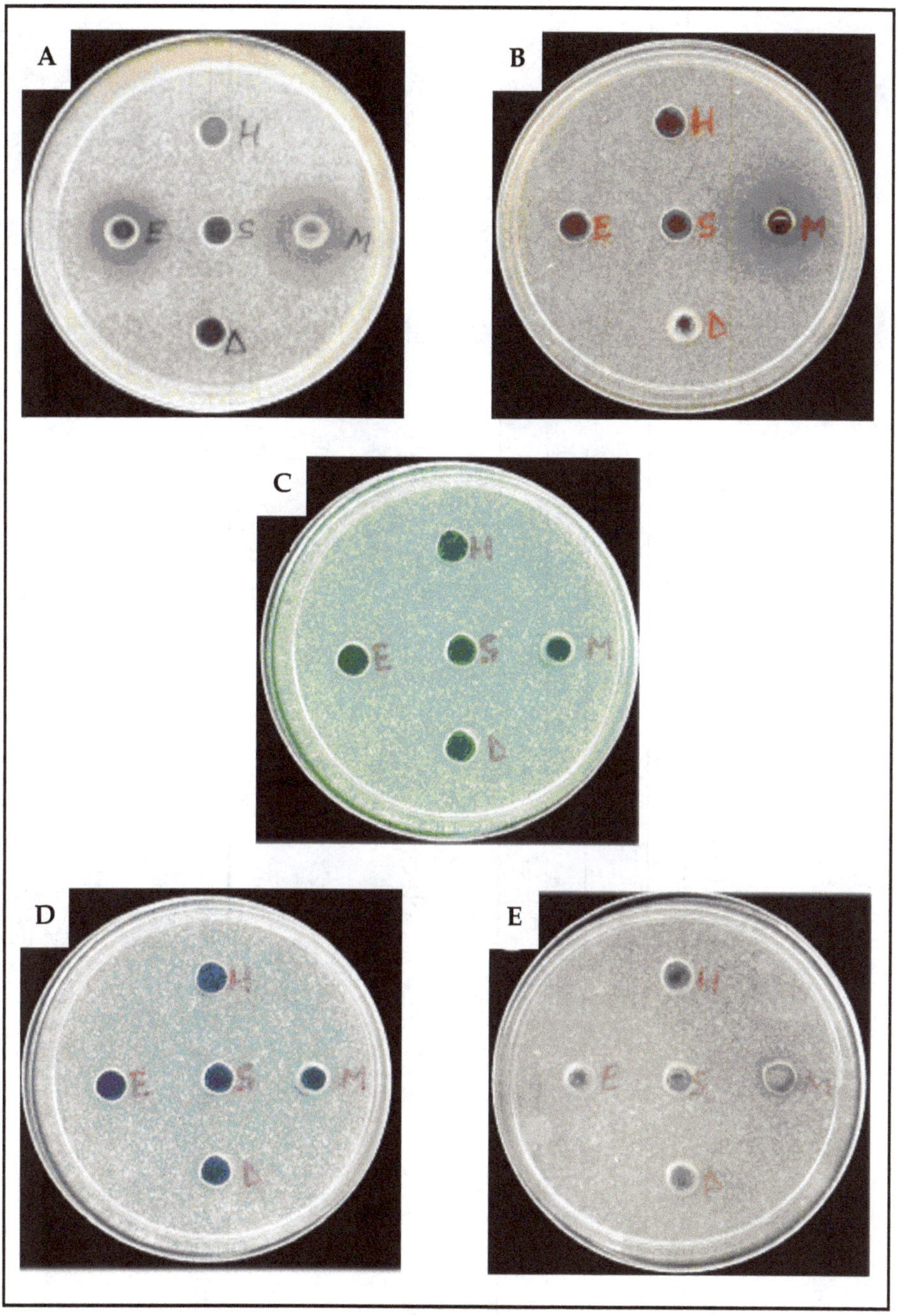

Fig. 39. Zone of inhibition exhibited by different extracts of *Nostoc linckia*. **A.** *S. aureus*; **B.** *E. coli*. **C.** *P aeruginosa*; **D.** *K. pneumoniae*; **E.** *S. typhi*. **Inside Petriplates**. H, hexane extract ; E, ethyl acetate extract; D, dichloromethane extract; M, methanolic extract; S, spent medium

Table 19. Antibacterial activity of different extracts of *Nostoc commune*

Test Organisms	Effective zone of inhibition (mm)										
	Organic Extracts (10mg/ml)								Extracellular substances		
	Hexane		Ethyl acetate		Dichloromethane		Methanol		Spent medium	BG-11 Media Control	Ciprofloxacin (5 μg/ml)
	Extract	Control	Extract	Control	Extract	Control	Extract	Control			
Staphylococcus aureus	0.0	0.0	6.79±.53	0.0	0.0	0.0	10.42±.6	0.0	0.0	0.0	25.33±.15
Escherichia coli	0.0	0.0	0.0	0.0	0.0	0.0	7.56±.7	0.0	0.0	0.0	23.46±.4
Pseudomonas aeruginosa	0.0	0.0	0.0	0.0	0.0	0.0	0.0	0.0	0.0	0.0	21.35±.22
Klebsiella pneumoniae	0.0	0.0	0.0	0.0	0.0	0.0	0.0	0.0	0.0	0.0	17.35±.22
Salmonella typhi	0.0	0.0	0.0	0.0	0.0	0.0	0.0	0.0	0.0	0.0	28.44±.28

Results are the means of diameter values ± standard deviation. Effective zone of inhibition (total zone of inhibition, minus diameter of well)

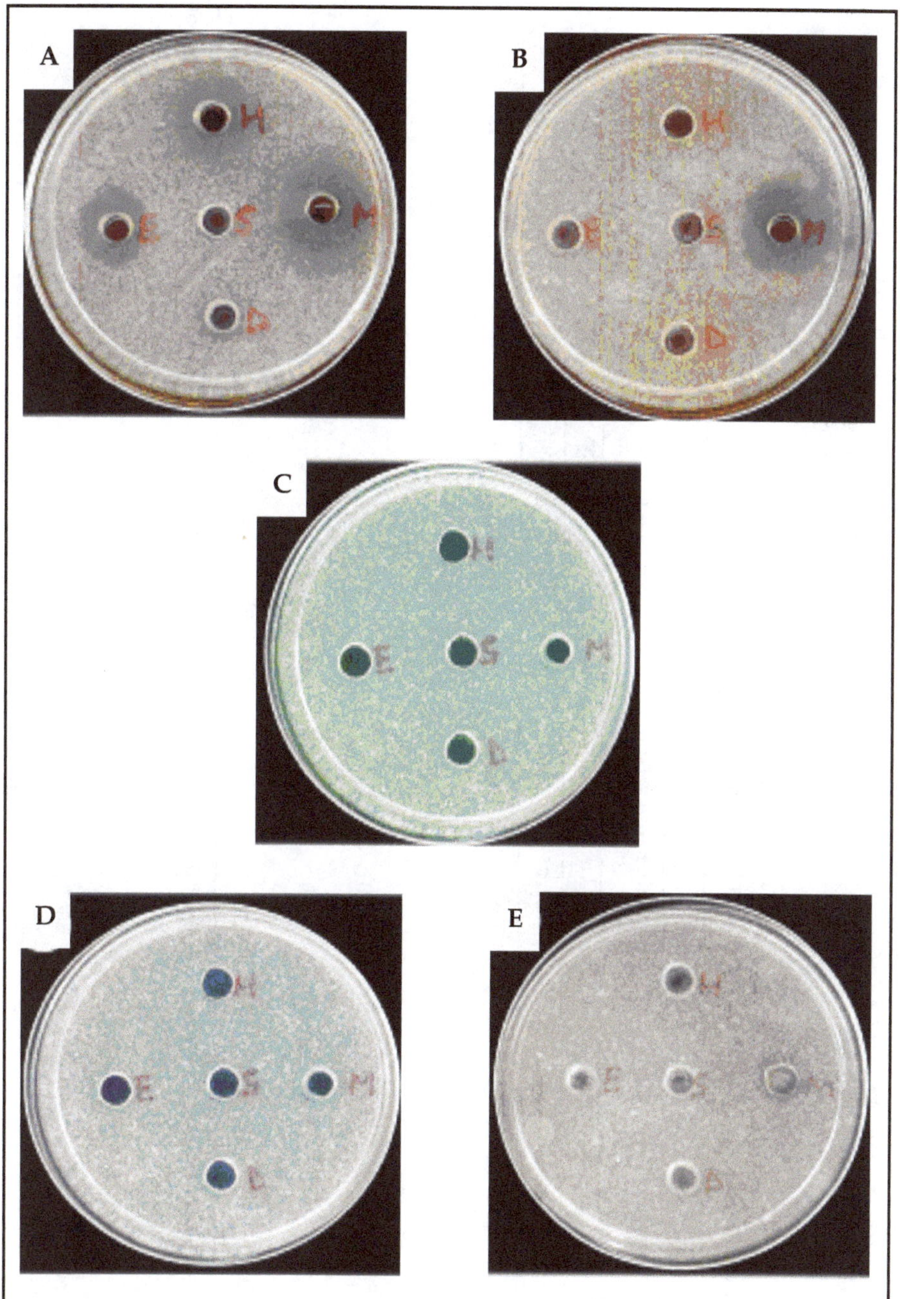

Fig. 40. Zone of inhibition exhibited by different extracts of *Nostoc commune*. **A.** *S. aureus*; **B.** *E. coli*; **C.** *P. aeruginosa*; **D.** *K. pneumoniae*; **E.** *S. typhi*. **Inside Petriplates**: H, hexane extract ; E, ethyl acetate extract; D, dichloromethane extract; M, methanolic extract; S, spent medium

Table 20. Antibacterial activity of different extracts of *Spirulina platensis*.

Test organisms	Effective zone of inhibition (mm)										
	Organic Extracts (10mg/ml)								Extracellular substances		
	Hexane		Ethyl acetate		Dichloromethane		Methanol		Spent medium	BG-11 Media Control	Ciprofloxacin (5 μg/ml)
	Extract	Control	Extract	Control	Extract	Control	Extract	Control			
Staphylococcus aureus	8.29±.2	0.0	4.45±1.3	0.0	11.38±.18	0.0	15.21±.1	0.0	0.0	0.0	25.33±.15
Escherichia coli	0.0	0.0	0.0	0.0	0.0	0.0	12.42±.47	0.0	7.26±.1	0.0	23.46±.4
Pseudomonas aeruginosa	0.0	0.0	0.0	0.0	0.0	0.0	11.55±.38	0.0	0.0	0.0	21.35±.22
Klebsiella pneumoniae	0.0	0.0	0.0	0.0	0.0	0.0	0.0	0.0	0.0	0.0	17.35±.22
Salmonella typhi	0.0	0.0	0.0	0.0	0.0	0.0	11.52±.18	0.0	0.0	0.0	28.44±.28

Results are the means of diameter values ± standard deviation. Effective zone of inhibition (total zone of inhibition, minus diameter of well)

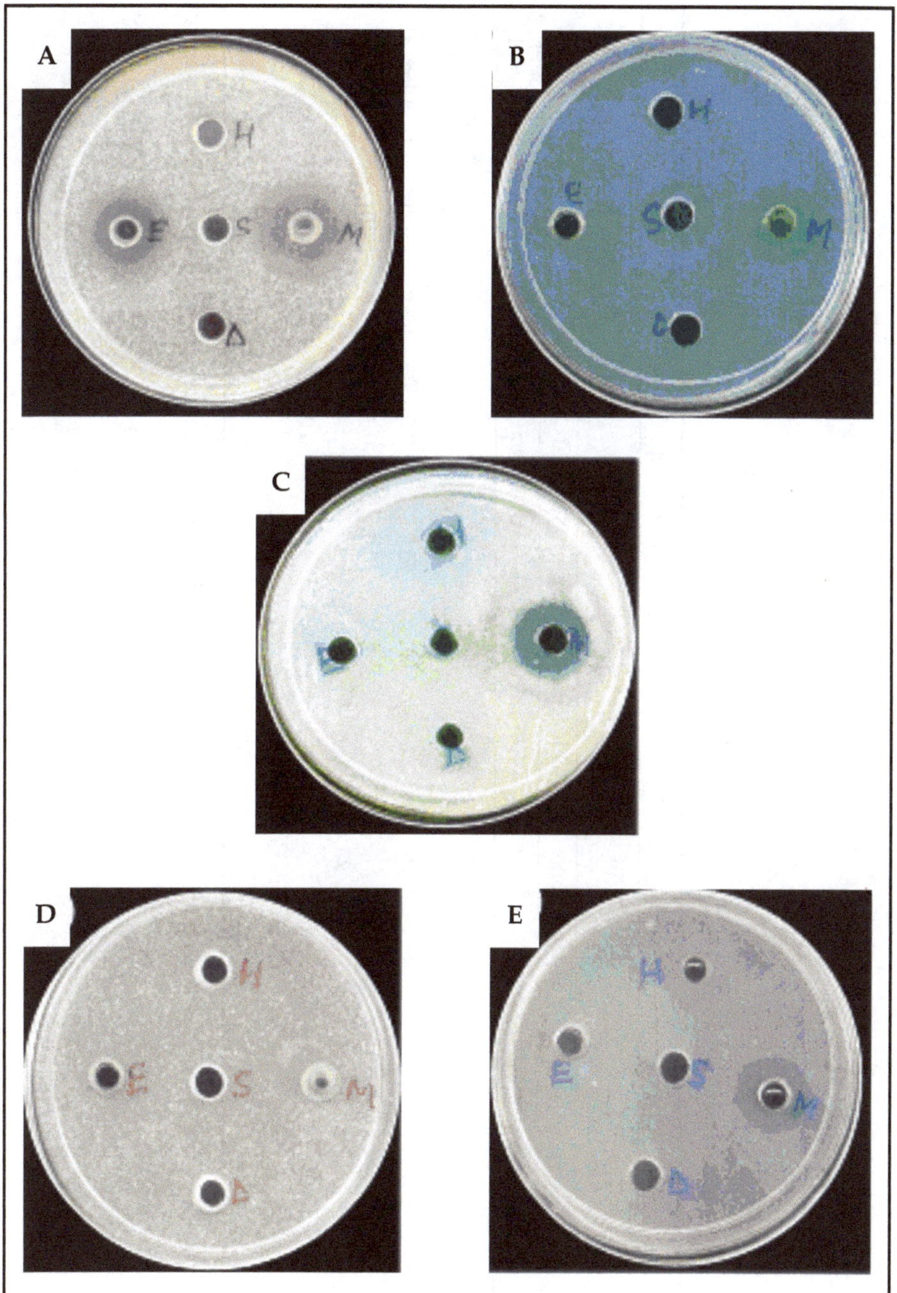

Fig. 41. Zone of inhibition exhibited by different extracts of *Spirulina platensis*. A. *S. aureus*; B. *E. coli*; C. *P. aeruginosa*; D. *K. pneumoniae*; E. *S. typhi*. Inside Petriplates: H, hexane extract ; E, ethyl acetate extract; D, dichloromethane extract; M, methanolic extract; S, spent medium

Table 21. Antibacterial activity of different extracts of *Westiellopsis prolifica*

Test organisms	Effective zone of inhibition (mm)										
	Organic Extracts (10mg/ml)								Extracellular substances		
	Hexane		Ethyl acetate		Dichloromethane		Methanol		Spent medium	BG-11 Media Control	Ciprofloxacin (5 µg/ml)
	Extract	Control	Extract	Control	Extract	Control	Extract	Control			
Staphylococcus aureus	4.45±1.3	0.0	4.37±1.4	0.0	5.39±.66	0.0	8.53±1.4	0.0	0.0	0.0	25.33±.15
Escherichia coli	0.0	0.0	0.0	0.0	0.0	0.0	0.0	0.0	0.0	0.0	23.46±.4
Pseudomonas aeruginosa	0.0	0.0	0.0	0.0	0.0	0.0	0.0	0.0	0.0	0.0	21.35±.22
Klebsiella pneumoniae	0.0	0.0	0.0	0.0	0.0	0.0	0.0	0.0	0.0	0.0	17.35±.22
Salmonella typhi	0.0	0.0	0.0	0.0	0.0	0.0	0.0	0.0	0.0	0.0	28.44±.28

Results are the means of diameter values ± standard deviation. Effective zone of inhibition (total zone of inhibition, minus diameter of well)

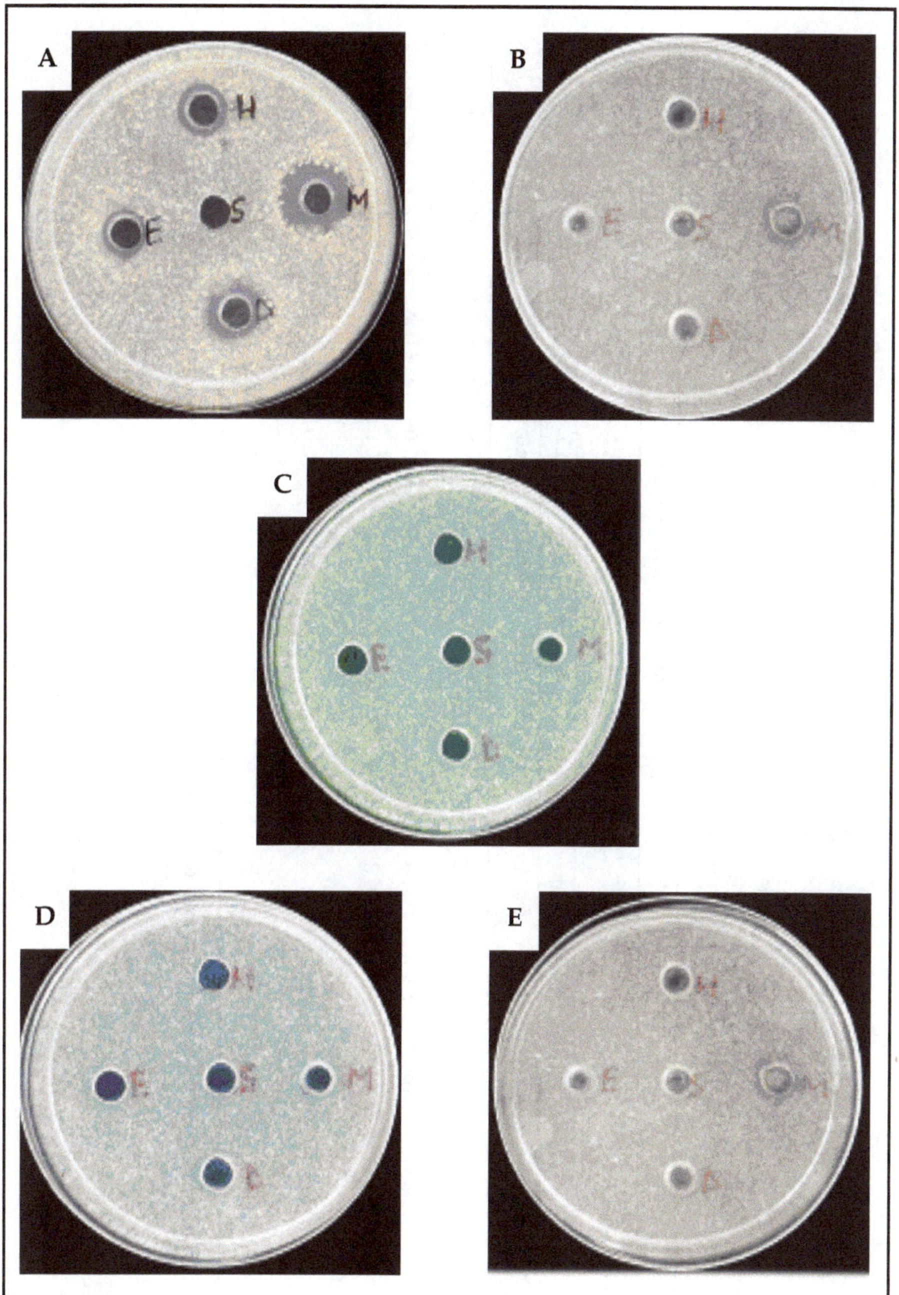

Fig. 42. Zone of inhibition exhibited by different extracts of *Westiellopsis prolifica*. A. *S. aureus*; B. *E. coli*; C. *P. aeruginosa*; D. *K. pneumoniae*; E. *S. typhi*. Inside Petriplates: H, hexane extract ; E, ethyl acetate extract; D, dichloromethane extract; M, methanolic extract; S, spent medium

Table 22. Antibacterial activity of different extracts of *Hapalosiphon sp.*

Test organisms	Effective zone of inhibition (mm)										
	Organic Extracts (10mg/ml)								Extracellular substances		Ciprofloxacin (5 μg/ml)
	Hexane		Ethyl acetate		Dichloromethane		Methanol		Spent medium	BG-11 Media Control	
	Extract	Control	Extract	Control	Extract	Control	Extract	Control			
Staphylococcus aureus	0.0	0.0	0.0	0.0	0.0	0.0	8.42±1.1	0.0	0.0	0.0	25.33±.15
Escherichia coli	0.0	0.0	0.0	0.0	0.0	0.0	5..74±.85	0.0	0.0	0.0	23.46±.4
Pseudomonas aeruginosa	0.0	0.0	0.0	0.0	0.0	0.0	0.0	0.0	0.0	0.0	21.35±.22
Klebsiella pneumoniae	0.0	0.0	0.0	0.0	0.0	0.0	0.0	0.0	0.0	0.0	17.35±.22
Salmonella typhi	0.0	0.0	0.0	0.0	0.0	0.0	0.0	0.0	0.0	0.0	28.44±.28

Results are the means of diameter values ± standard deviation. Effective zone of inhibition (total zone of inhibition, minus diameter of well)

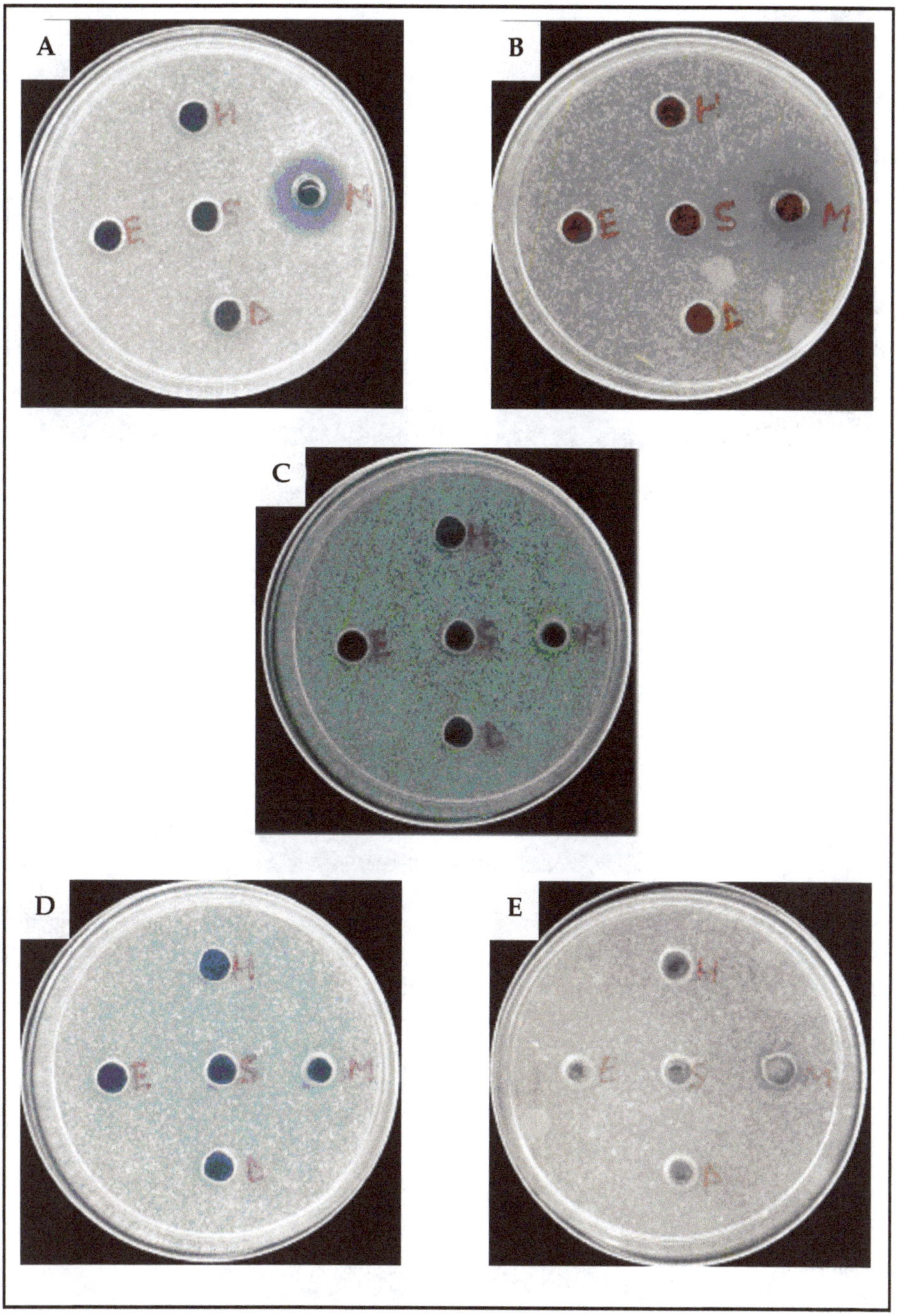

Fig. 43. Zone of inhibition exhibited by different extracts of *Hapalosiphon* sp.; **A.** *S. aureus*; **B.** *E. coli*; **C.** *P. aeruginosa*; **D.** *K. pneumoniae*; **E.** *S. typhi*. **Inside Petriplates:** H, hexane extract ; E, ethyl acetate extract; D, dichloromethane extract; M, methanolic extract; S, spent medium

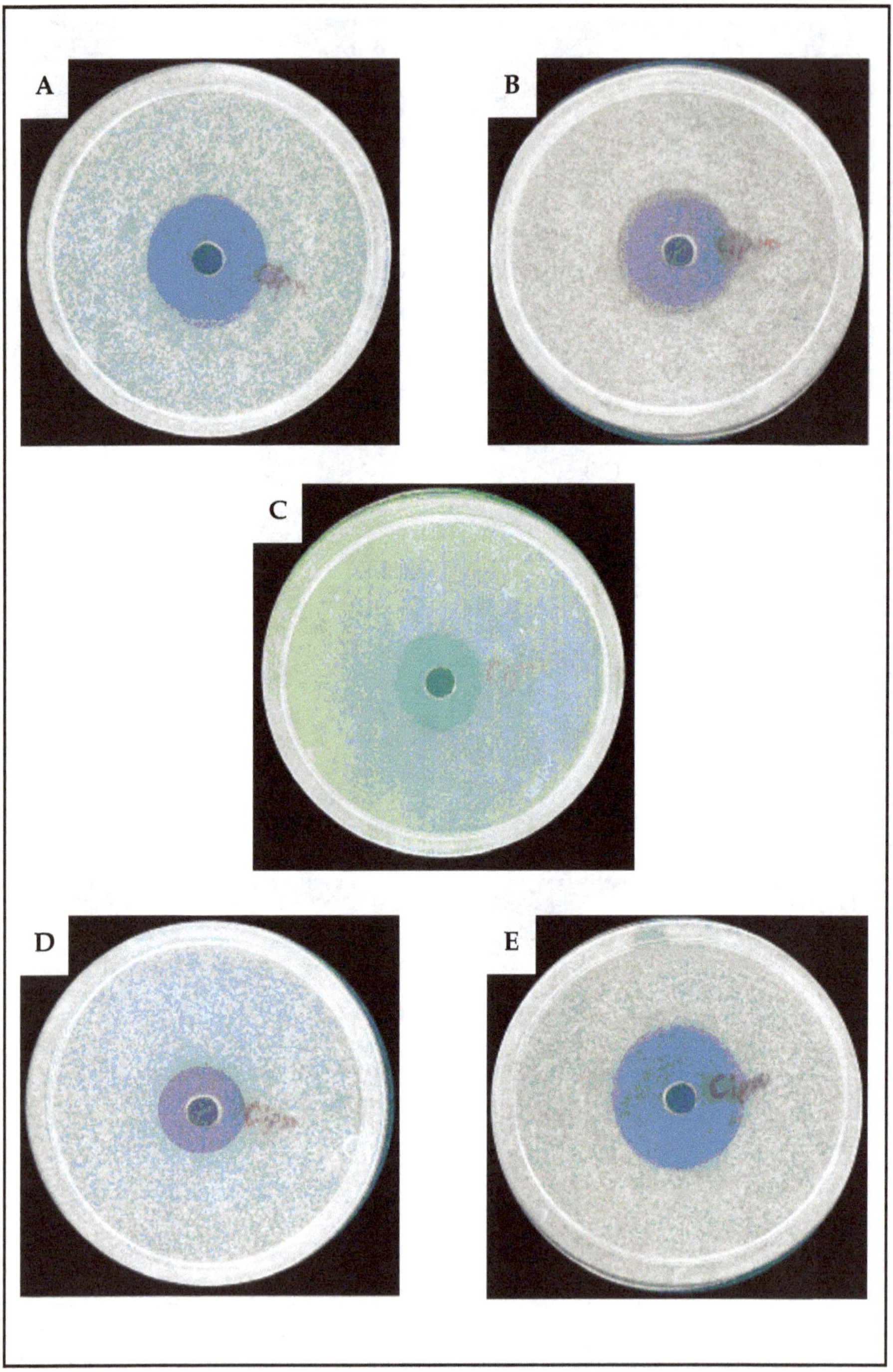

Fig. 44. Zone of inhibition exhibited by standard antibiotic *Ciprofloxacin*. **A.** *S. aureus*; **B.** *E. coli*; **C.** *P. aeruginosa*; **D.** *K. pneumoniae*; **E.** *S. typhi*.

Table 23. Minimum Inhibitory Concentration of active crude extracts of *Anabaena variabilis*

Extracts	Test organisms	Concentration of extracts in µg/ml										MIC
		512	256	128	64	32	16	8	4	2	1	
Hexane	*S. aureus*	+	+	+	+	+	+	+	+	+	+	>512
Ethyl acetate	*S. aureus*	-	+	+	+	+	+	+	+	+	+	512
Ethyl acetate	*E. coli*	+	+	+	+	+	+	+	+	+	+	>512
Methanol	*S. aureus*	-	-	+	+	+	+	+	+	+	+	256
Methanol	*E. coli*	-	+	+	+	+	+	+	+	+	+	512
Methanol	*P. aeruginosa*	+	+	+	+	+	+	+	+	+	+	>512
Methanol	*S. typhi*	+	+	+	+	+	+	+	+	+	+	>512
Ciprofloxacin	*S. aureus*	-	-	-	-	-	-	-	-	-	-	<1
Ciprofloxacin	*E. coli*	-	-	-	-	-	-	-	-	-	-	<1
Ciprofloxacin	*P. aeruginosa*	-	-	-	-	-	-	-	-	-	-	<1
Ciprofloxacin	*S. typhi*	-	-	-	-	-	-	-	-	-	-	<1

(-) No growth observed, (+) Growth observed

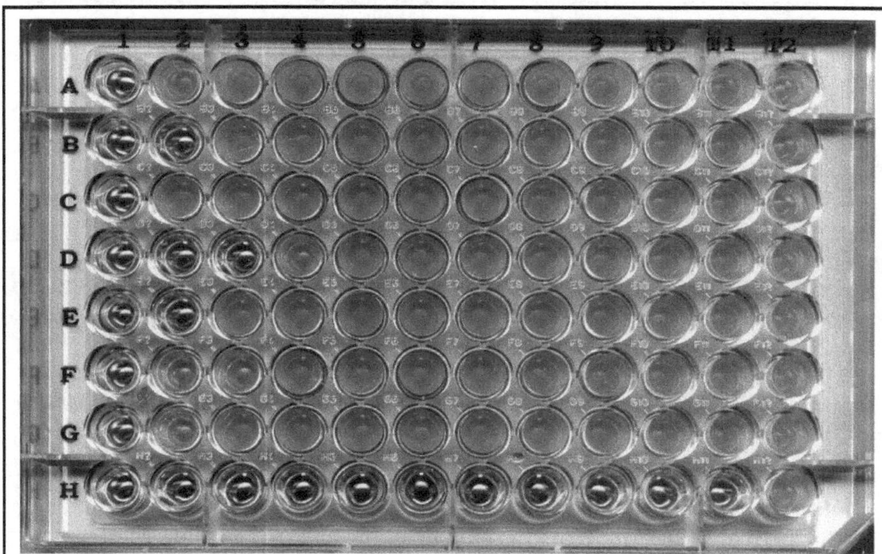

Fig. 45. MIC analysis for *Anabaena variabilis* extracts: A ninty six well microtiter plate showing the inhibition of growth

Rows : A. Hexane extracts against *S. aureus* **B.** Ethyl acetate extract against *S. aureus* **C.** Ethyl acetate extract against *E. coli* **D.** Methanolic extract against *S. aureus* **E.** Methanolic extract against *E. coli* **F.** Methanolic extract against *P. aeruginosa* **G.** Methanolic extract against *S. typhi* **H.** Ciprofloxacin against *S. aureus*

Column : 1. Negative control (MHB) **2.** 11-Dilution range 512 µg/ml to 1 µg/ml. **3.** 512 µg/ml, 3. 256 µg/ml, **4.** 128 µg/ml **5.** 64 µg/ml, **6.** 32 µg/ml, **7.** 16 µg/ml, **8.** 8 µg/ml **9.** 4 µg/ml, **10.** 2 µg/ml **11.** 1 µg/ml **12.** Positive control (MHB + Culture)

Table 24. Minimum Inhibitory Concentration of active crude extracts of *Anabaena fertilissima*

Extracts	Test organisms	Concentration of extracts in µg/ml										MIC
		512	256	128	64	32	16	8	4	2	1	
Ethyl acetate	*S. aureus*	+	+	+	+	+	+	+	+	+	+	>512
Ethyl acetate	*E. coli*	+	+	+	+	+	+	+	+	+	+	>512
Methanol	*S. aureus*	-	+	+	+	+	+	+	+	+	+	512
Methanol	*E. coli*	-	+	+	+	+	+	+	+	+	+	512
Methanol	*P. aeruginosa*	+	+	+	+	+	+	+	+	+	+	>512
Methanol	*S. typhi*	+	+	+	+	+	+	+	+	+	+	>512
Ciprofloxacin	*S. aureus*	-	-	-	-	-	-	-	-	-	-	<1
Ciprofloxacin	*E. coli*	-	-	-	-	-	-	-	-	-	-	<1
Ciprofloxacin	*P. aeruginosa*	-	-	-	-	-	-	-	-	-	-	<1
Ciprofloxacin	*S. typhi*	-	-	-	-	-	-	-	-	-	-	<1

(-) No growth observed, (+) Growth observed

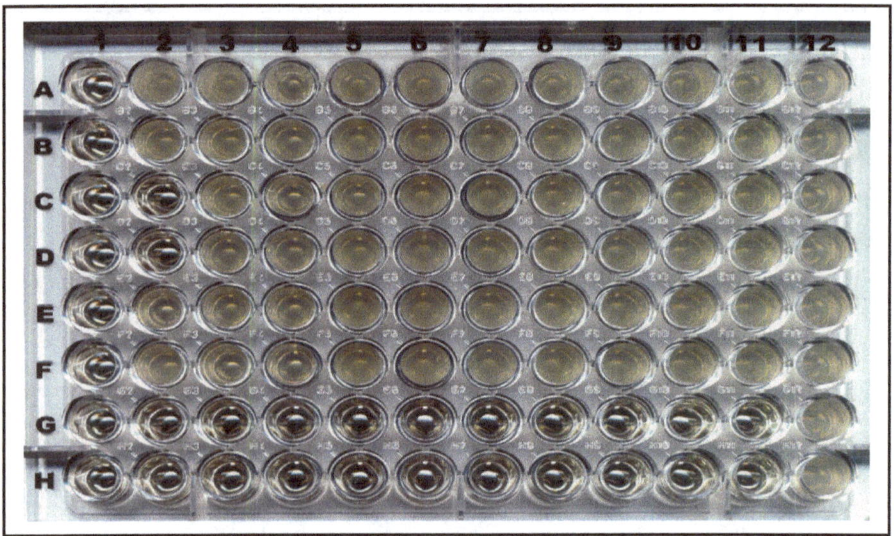

Fig. 46. MIC analysis for *Anabaena fertilissima* extracts: A ninty six well microtiter plate showing the inhibition of growth

Rows : A. Ethyl acetate extract against *S. aureus*
B. Ethyl acetate extract against *E. coli* **C.** Methanolic extract against *S. aureus*
D. Methanolic extract against *E. coli* **E.** Methanolic extract against *P. aeruginosa*
F. Methanolic extract against *S. typhi* **G.** Ciprofloxacin against *S. aureus*
H. Ciprofloxacin against *E. coli*

Column : 1. Negative control (MHB) **2.** 11-Dilution range 512 µg/ml to 1 µg/ml.
2. 512 µg/ml, **3.** 256 µg/ml, **4.** 128 µg/ml **5.** 64 µg/ml, **6.** 32 µg/ml, **7.** 16 µg/ml,
8. 8 µg/ml **9.** 4 µg/ml, **10.** 2 µg/ml, **11.** 1 µg/ml **12.** Positive control (MHB + Culture)

Table 25. Minimum Inhibitory Concentration of active crude extracts of *Nostoc muscorum*

Extracts	Test organisms	Concentration of extracts in µg/ml										MIC
		512	256	128	64	32	16	8	4	2	1	
Hexane	*S. aureus*	+	+	+	+	+	+	+	+	+	+	>512
Ethyl acetate	*S. aureus*	-	+	+	+	+	+	+	+	+	+	512
Ethyl acetate	*E. coli*	-	+	+	+	+	+	+	+	+	+	512
Methanol	*S. aureus*	-	-	-	+	+	+	+	+	+	+	128
Methanol	*E. coli*	-	-	+	+	+	+	+	+	+	+	256
Methanol	*P. aeruginosa*	+	+	+	+	+	+	+	+	+	+	>512
Methanol	*S. typhi*	+	+	+	+	+	+	+	+	+	+	>512
Ciprofloxacin	*S. aureus*	-	-	-	-	-	-	-	-	-	-	<1
Ciprofloxacin	*E. coli*	-	-	-	-	-	-	-	-	-	-	<1
Ciprofloxacin	*P. aeruginosa*	-	-	-	-	-	-	-	-	-	-	<1
Ciprofloxacin	*S. typhi*	-	-	-	-	-	-	-	-	-	-	<1

(-) No growth observed, (+) Growth observed

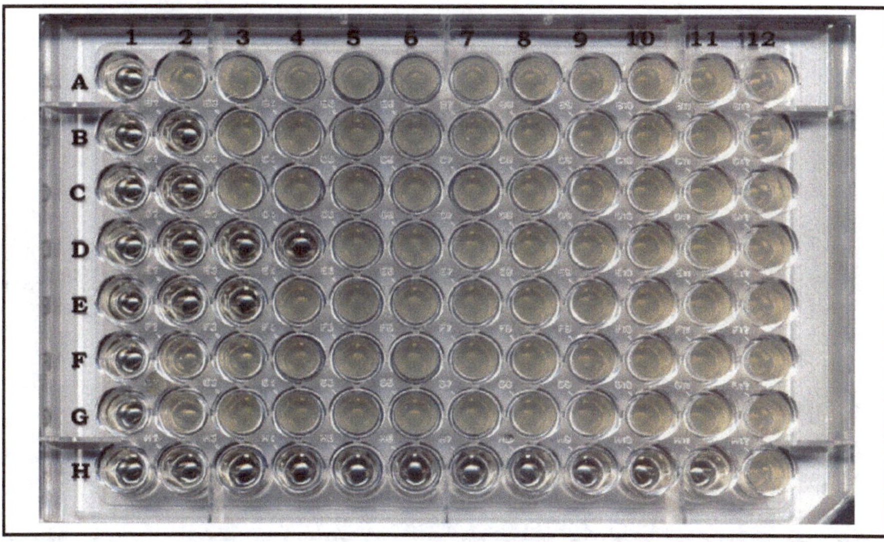

Fig. 47. MIC analysis for *Nostoc muscorum* extracts: A ninty six well microtiter plate showing the inhibition of growth

Rows : A. Hexane extracts against *S. aureus* **B.** Ethyl acetate extract against *S. aureus* **C.** Ethyl acetate extract against *E. coli* **D.** Methanolic extract against *S. aureus* **E.** Methanolic extract against *E. coli* **F.** Methanolic extract against *P. aeruginosa* **G.** Methanolic extract against *S. typhi* **H.** Ciprofloxacin against *S. aureus*

Column : 1. Negative control (MHB) **2.** 11-Dilution range 512 µg/ml to 1 µg/ml. **2.** 512 µg/ml, **3.** 256 µg/ml, **4.** 128 µg/ml **5.** 64 µg/ml, **6.** 32 µg/ml, **7.** 16 µg/ml, **8.** 8 µg/ml **9.** 4 µg/ml, **10.** 2 µg/ml, **11.** 1 µg/ml **12.** Positive control (MHB + Culture)

Table 26. Minimum Inhibitory Concentration of active crude extracts of *Nostoc punctiforme*

Extracts	Test organisms	Concentration of extracts in μg/ml										MIC
		512	256	128	64	32	16	8	4	2	1	
Ethyl acetate	*S. aureus*	-	+	+	+	+	+	+	+	+	+	512
Ethyl acetate	*S. aureus*	+	+	+	+	+	+	+	+	+	+	>512
Ethyl acetate	*E. coli*	+	+	+	+	+	+	+	+	+	+	>512
Methanol	*S. aureus*	-	+	+	+	+	+	+	+	+	+	512
Methanol	*E. coli*	-	+	+	+	+	+	+	+	+	+	512
Methanol	*P. aeruginosa*	+	+	+	+	+	+	+	+	+	+	>512
Methanol	*S. typhi*	+	+	+	+	+	+	+	+	+	+	>512
Ciprofloxacin	*S. aureus*	-	-	-	-	-	-	-	-	-	-	<1
Ciprofloxacin	*E. coli*	-	-	-	-	-	-	-	-	-	-	<1
Ciprofloxacin	*P. aeruginosa*	-	-	-	-	-	-	-	-	-	-	<1
Ciprofloxacin	*S. typhi*	-	-	-	-	-	-	-	-	-	-	<1

(-) No growth observed, (+) Growth observed

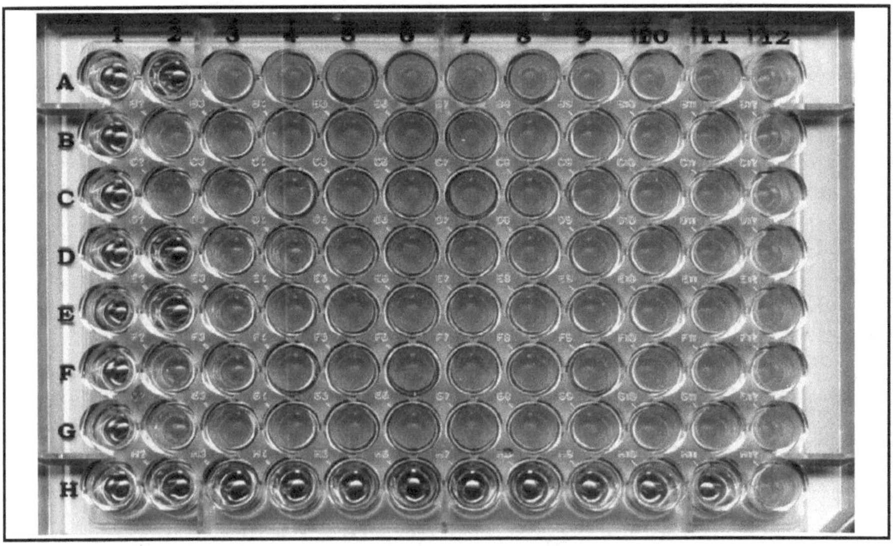

Fig. 48. MIC analysis for *Nostoc muscorum* extracts: A ninty six well microtiter plate showing the inhibition of growth

Rows : A. Ethyl acetate extract against *S. aureus* **B.** Ethyl acetate extract against *E. coli* **C.** Ethyl acetate extract against *P. aeruginosa* **D.** Methanolic extract against *S. aureus* **E.** Methanolic extract against *E. coli* **F.** Methanolic extract against *P. aeruginosa* **G.** Methanolic extract against *S. typhi* **H.** Ciprofloxacin against *S. aureus*
Column : 1. Negative control (MHB) **2.** 11-Dilution range 512 μg/ml to 1 μg/ml. **2.** 512 μg/ml, **3.** 256 μg/ml, **4.** 128 μg/ml **5.** 64 μg/ml, **6.** 32 μg/ml, **7.** 16 μg/ml, **8.** 8 μg/ml **9.** 4 μg/ml, **10.** 2 μg/ml, **11.** 1 μg/ml **12.** Positive control (MHB + Culture)

Table 27. Minimum Inhibitory Concentration of active crude extracts of *Nostoc linckia.*

Extracts	Test organisms	Concentration of extracts in µg/ml										MIC
		512	256	128	64	32	16	8	4	2	1	
Ethyl acetate	*S. aureus*	+	+	+	+	+	+	+	+	+	+	>512
Methanol	*S. aureus*	+	+	+	+	+	+	+	+	+	+	>512
Methanol	*E.coli*	+	+	+	+	+	+	+	+	+	+	>512
Ciprofloxacin	*S. aureus*	-	-	-	-	-	-	-	-	-	-	<1
Ciprofloxacin	*E.coli*	-	-	-	-	-	-	-	-	-	-	<1

(-) No growth observed, (+) Growth observed,

Table 28. Minimum Inhibitory Concentration of active crude extracts of *Nostoc commune*

Extracts	Test organisms	Concentration of extracts in µg/ml										MIC
		512	256	128	64	32	16	8	4	2	1	
Ethyl acetate	*S. aureus*	+	+	+	+	+	+	+	+	+	+	>512
Methanol	*S. aureus*	+	+	+	+	+	+	+	+	+	+	>512
Methanol	*E.coli*	+	+	+	+	+	+	+	+	+	+	>512
Ciprofloxacin	*S. aureus*	-	-	-	-	-	-	-	-	-	-	<1
Ciprofloxacin	*E.coli*	-	-	-	-	-	-	-	-	-	-	<1

(-) No growth observed, (+) Growth observed

Table 29. Minimum Inhibitory Concentration of active crude extracts of *Spirulina platensis*

Extracts	Test organisms	Concentration of extracts in µg/ml										MIC
		512	256	128	64	32	16	8	4	2	1	
Hexane	*S. aureus*	+	+	+	+	+	+	+	+	+	+	>512
Ethyl acetate	*S. aureus*	+	+	+	+	+	+	+	+	+	+	>512
Dichlkoromethane	*E.coli*	-	+	+	+	+	+	+	+	+	+	512
Methanol	*S. aureus*	-	-	-	+	+	+	+	+	+	+	128
Methanol	*E.coli*	-	-	+	+	+	+	+	+	+	+	256
Methanol	*P.aeruginosa*	+	+	+	+	+	+	+	+	+	+	>512
Methanol	*S.typhi*	+	+	+	+	+	+	+	+	+	+	>512
Ciprofloxacin	*S. aureus*	-	-	-	-	-	-	-	-	-	-	<1
Ciprofloxacin	*E.coli*	-	-	-	-	-	-	-	-	-	-	<1
Ciprofloxacin	*P.aeruginosa*	-	-	-	-	-	-	-	-	-	-	<1
Ciprofloxacin	*S.typhi*	-	-	-	-	-	-	-	-	-	-	<1

(-) No growth observed, (+) Growth observed

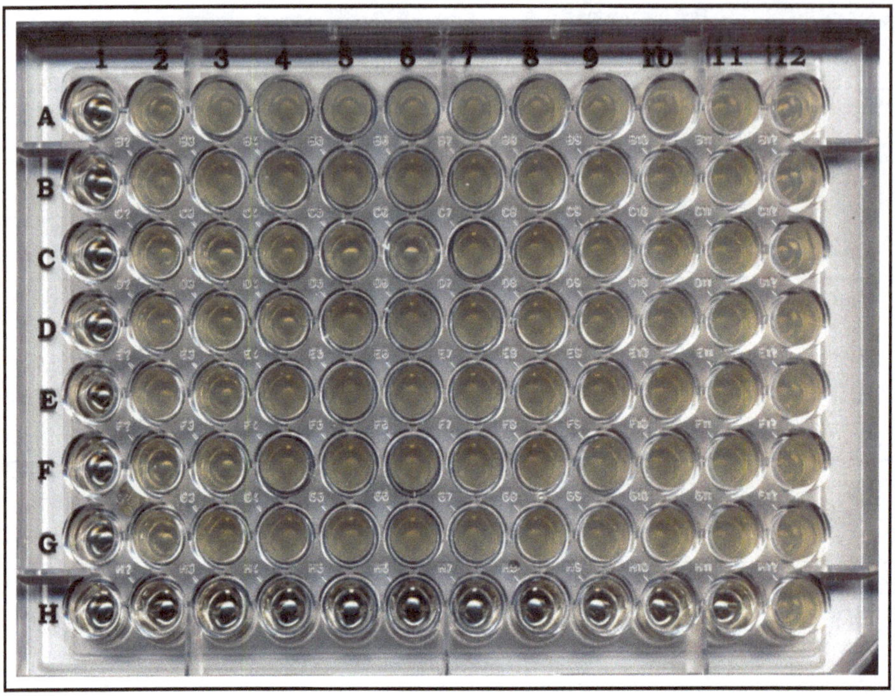

Fig. 49. MIC analysis for *Nostoc linckia* and *Nostoc commune* extracts: A ninty six well microtiter plate showing the inhibition of growth

Rows : A. Ethyl acetate extract against *Nostoc linckia* against *S. aureus*

B. Methanol extract of *Nostoc linckia* against *S. aureus*

C. Methanol extract of *Nostoc linckia* against *E. coli*

D. Ethyl acetate extract of *Nostoc commune* against *S. aureus*

E. Methanolic extract of *Nostoc commune* against *S. aureus*

F. Methanolic extract of *Nostoc commune* against *E. coli*

G. Ciprogloxacin against *S. aureus*

H. Ciprofloxacin against *E. coli*

Column : 1. Negative control (MHB) **2.** 11-Dilution range 512 μg/ml to 1 μg/ml. **2.** 512 μg/ml, **3.** 256 μg/ml, **4.** 128 μg/ml **5.** 64 μg/ml, **6.** 32 μg/ml, **7.** 16 μg/ml, **8.** 8 μg/ml **9.** 4 μg/ml, **10.** 2 μg/ml, **11.** 1 μg/ml **12.** Positive control (MHB + Culture)

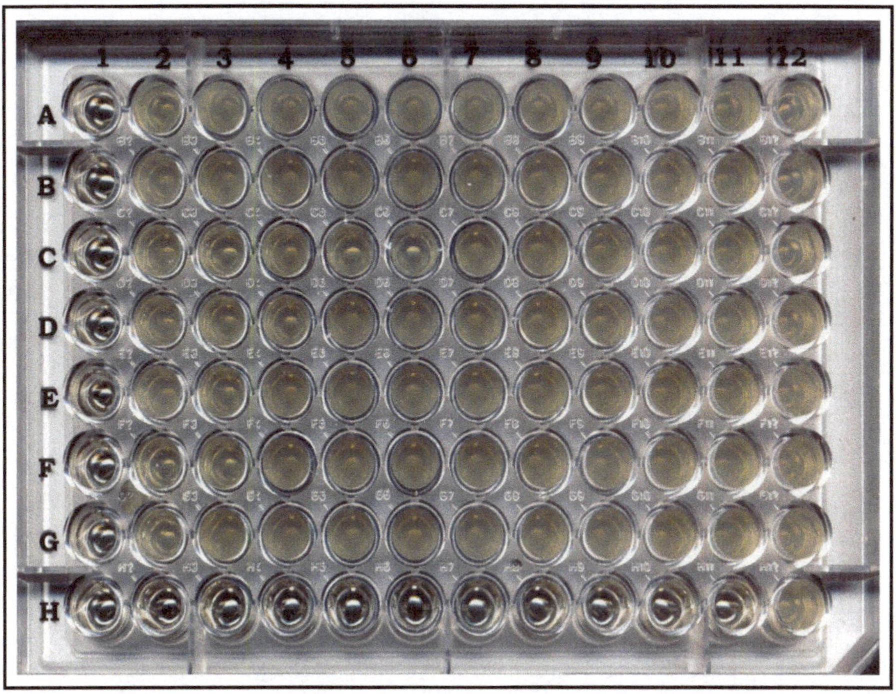

Fig. 50. MIC analysis for *Spirulina platensis* extracts: A ninty six well microtiter plate showing the inhibition of growth

Rows : A. Hexane extracts against *S. aureus*

B. Ethyl acetate extract against *S. aureus*

C. Di chloro methane extract against *S. aureus*

D. Methanolic extract against *S. aureus*

E. Methanolic extract against *E. coli*

F. Methanolic extract against *P. aeruginosa*

G. Methanolic extract against *S. typhi*

H. Ciprofloxacin against *S. aureus*

Column : 1. Negative control (MHB) **2.** 11-Dilution range 512 μg/ml to 1 μg/ml.
2. 512 μg/ml, **3.** 256 μg/ml, **4.** 128 μg/ml **5.** 64 μg/ml, **6.** 32 μg/ml, **7.** 16 μg/ml,
8. 8 μg/ml **9.** 4 μg/ml, **10.** 2 μg/ml, **11.** 1 μg/ml **12.** Positive control (MHB + Culture)

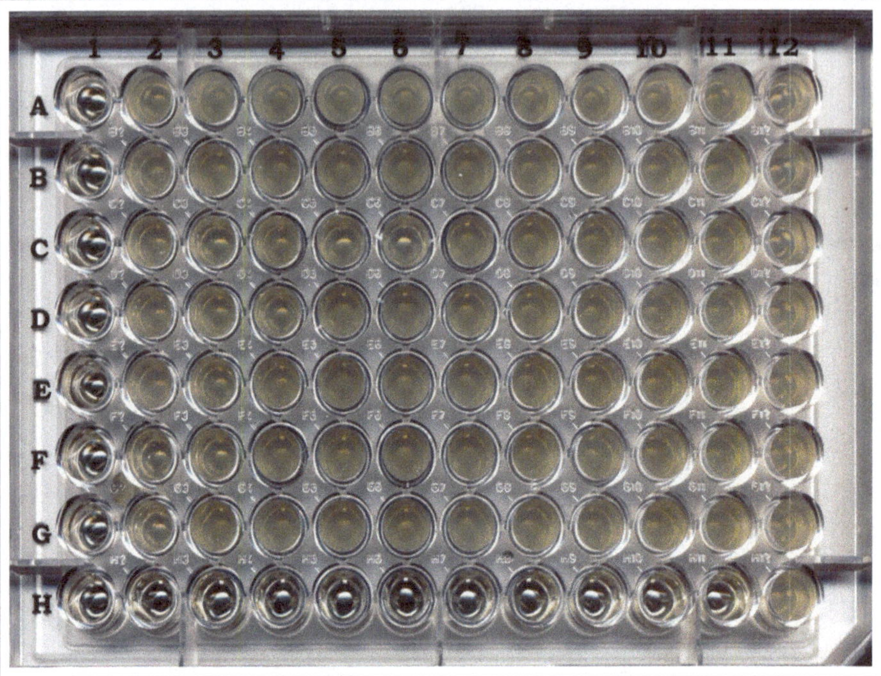

Fig. 51. MIC analysis for *Westiellopsis prolifica* and *Hapalosiphon* sp. extracts. A ninty six well microtiter plate showing the inhinition of growth

Rows : A. Hexane extracts of *Westiellopsis prolifica* against *S. aureus*

B. Ethyl acetate extract of *Westiellopsis prolifica* against *S. aureus*

C. Di chloro methane extract of *Westiellopsis prolifica* against *S. aureus*

D. Methanolic extract of *Westiellopsis prolifica* against *S. aureus*

E. Methanolic extract of *Hapalosiphon sp.* against *S. aureus*

F. Methanolic extract of *Hapalosiphon sp.* against *E. coli*

G. Ciprofloxacin against *S. aureus*

H. Ciprofloxacin against *E. coli*

Column : 1. Negative control (MHB) **2.** 11-Dilution range 512 μg/ml to 1 μg/ml. **2.** 512 μg/ml, **3.** 256 μg/ml, **4.** 128 μg/ml **5.** 64 μg/ml, **6.** 32 μg/ml, **7.** 16 μg/ml, **8.** 8 μg/ml **9.** 4 μg/ml, **10.** 2 μg/ml, **11.** 1 μg/ml **12.** Positive control (MHB + Culture)

Table 30. Minimum Inhibitory Concentration of active crude extracts of *Westiellopsis prolifica*

Extracts	Test organisms	Concentration of extracts in µg/ml										MIC
		512	256	128	64	32	16	8	4	2	1	
Hexane	*S. aureus*	+	+	+	+	+	+	+	+	+	+	>512
Ethyl acetate	*S. aureus*	+	+	+	+	+	+	+	+	+	+	>512
Dichloromethane	*S aureus*	+	+	+	+	+	+	+	+	+	+	>512
Methanol	*S. aureus*	+	+	+	+	+	+	+	+	+	+	>512
Ciprofloxacin	*S. aureus*	-	-	-	-	-	-	-	-	-	-	<1

(-) No growth observed, (+) Growth observed

Table 31. Minimum Inhibitory Concentration of active crude extracts of *Hapalosiphon sp.*

Extracts	Test organisms	Concentration of extracts in µg/ml										MIC
		512	256	128	64	32	16	8	4	2	1	
Methanol	*S. aureus*	+	+	+	+	+	+	+	+	+	+	>512
Methanol	*E. coli*	+	+	+	+	+	+	+	+	+	+	>512
Ciprofloxacin	*S. aureus*	-	-	-	-	-	-	-	-	-	-	<1
Ciprofloxacin	*E. coli*	-	-	-	-	-	-	-	-	-	-	<1

(-) No growth observed, (+) Growth observed

Chapter 6

DISCUSSION

In the present investigation, nine cyanobacterial species *viz. Anabaena variabilis, Anabaena fertilissima, Nostoc muscorum, Nostoc punctiforme, Nostoc linckia, Nostoc commune, Spirulina platensis, Westiellopsis prolifica* and *Hapalosiphon sp.* were tested for their activity pattern against some selected bacterial species.

Each cyanobacterial species was initially subjected to determine the early stationary phase of their life-cycle, which is believed to be the most appropriate period for the production of secondary metabolites by an organism. The growth curve for each cyanobacterial species was prepared by culturing them in BG-11 growth medium. Optical densities (OD) recorded at 650 nm were plotted against number of incubation days for each cyanobacterium. After 23-25 days of incubation period, each culture was harvested to obtain the biomass and spent medium, which were then subjected to process of extraction by using various solvents *viz.*, Hexane, ethyl acetate, dichloromethane and methanol. The extracts and spent medium thus obtained were then investigated for antibacterial activity against five pathogenic bacteria including *Staphylococcus aureus, Escherichia coli, Pseudomonas aeruginosa, Klebsiella pnumoniae* and *Salmonella typhi.*

Agar well diffusion assay was the key process to determine the antibacterial activity of all the extracts prepared in different solvents.

The antibacterial activity of *Anabaena variabilis* is shown in **Table 14.** Methanolic extract was found more active as compared to hexane, ethyl acetate and dichloromethane extracts. The highest activity in terms of effective zone of inhibition (10.14 mm) was observed against *S. aureus* followed by *E. coli, S. typhi* and *P. aeruginosa,* whereas no zone of inhibition was observed against *K. pneumoniae*. Hexane extracts showed little activity (4.31 mm) against *S. aureus*. Ethyl acetate extract was found active against *S. aureus* and *E. coli* with the effective zone of inhibition 6.42 mm and 5.41 mm, respectively. Dichloromethane extract and spent medium did not exhibit antibacterial activity .

Furthermore, the active extracts obtained from the above process *i.e.* hexane, ethyl acetate, and methanolic extracts were then subjected to the determination of minimum inhibitory concentrations by 96-well micro broth dilution method. Active crude extracts were diluted in CAMHB in the range from 512 μg/ml to 1 μg/ml using ciprofloxacin as a standard antibiotic. The results (Table 23) showed that the MIC of methanol extract was 256 μg/ml against *S. aureus* whereas it was 512 μg/ml against *E. coli,* However, it was found to be >512 μg/ml when tested against *P. aeruginosa* and *S. typhi* . The MIC of ethyl acetate extract was observed as 512 μg/ml against *S. aureus* but it was more than 512 μg/ml against *E. coli.* The MIC of hexane extract was recorded >512 μg/ml when tested against *S. aureus* .

The antibacterial activity of *Anabaena fertilissima* is summarized in **Table 15**. In this case, only two extracts *i.e.* methanolic and ethyl acetate extracts were found active as compared to previously studied *Anabaena variabilis*. The highest effective zone of inhibition (8.56 mm) had been observed in methanolic extract against *S. aureus* followed by *E. coli, P. aeruginosa* and *S. typhi*. Hexane extract, dichloromethane extract and spent medium were found inactive against all examined bacteria. *K. pneumoniae* was found completely resistant to all experimented extracts **(Table-15)**.

MIC values for the active extracts of *Anabaena fertilissima* are shown in Table 15, describing the methanolic extract having the MIC 512 μg/ml against *S. aureus* and *E. coli* , it was found to be >512 μg/ml when tested against *P. aeruginosa* and *S. typhi*. The MIC of ethyl extract was recorded more than 512 μg/ml against *S. aureus* and *E. coli*. (Table-24)

Mundt *et al.* (2001) found that the extract of *A. variabilis* inhibited the growth of only *Bacillus subtilis*. In contrast, the present results showed that different extracts of *A. variabilis* inhibited the growth of

S. aureus, E. coli, P. aeruginosa and *S. typhi*. Similarly, Falch *et al.* (1995) reported the broad spectrum antibacterial activity of different extracts of *Anabaena planctonica* 115b. Similarly Rao *et al.* (2007) reported the antimycobacterial activity of different extracts of *Anabaena sp., Anabaena variabiles, Spirulina sp., Anabaena cylindrica,* against *Mycobacterium tuberculosis* ATCC 27294, *M.tuberculosis MDR, M. avium, M. intracellulare* and *M. aurum.*

The antibacterial activity of different extracts of *Nostoc muscorum* against test bacteria is shown in **Table 16**. Methanolic extract of this cyanobacterium was found active against all the tested microorganisms except *K. pneumoniae*. The highest effective zone of inhibition (11.72 mm) was observed against *S. aureus* followed by 9.46 mm against *E. coli*, 8.45 mm against *P. aeruginosa* and only 5.09 mm against *S. typhi*. Hexane extracts showed little activity (3.32 mm) against *S. aureus*. Ethyl acetate extract was found active against *S. aureus* and *E. coli* with the effective zone of inhibition 8.50 mm and 7.33 mm, respectively. Dichloromethane extract and spent medium were found inactive against all the test bacteria.

The MIC values of *N. muscorum* have been shown in **Table 25**. The results showed that the MICs of methanolic extract were 128 µg/ml against *S. aureus* and 256 µg/ml against *E. coli* whereas it was more than 512 µg/ml against *P. aeruginosa* and *S. typhi*. The MIC of ethyl acetate extract was 512 µg/ml against S. *aureus* and *E. coli*. The MIC of hexane extract was found >512 ìg/ml against *S. aureus*.

The antibacterial activity of *Nostoc punctiforme* is summarized in **Table 8**. In this case, only methanolic and ethyl acetate extracts were found active. The highest effective zone of inhibition (7.58 mm) was observed in methanolic extract against *S. aureus* followed by *E. coli, P. aeruginosa* and *S. typhi*. Hexane extract, Dichloromethane extract and spent medium were found inactive against both types of bacteria. No activity was observed against *K. pneumoniae* **(Fig. 38)**.

The minimum inhibitory concentrations of the active extracts of *N. punctiforme* are shown in **Table 26** describing that the MICs of methanolic extract were 512 µg/ml against both *E. coli* and *S. aureus* whereas it was more than 512 µg/ml against *P. aeruginosa* and *S. typhi*. The MIC of ethyl acetate extracts was 512 µg/ml against *S. aureus* but it was >512 µg/ml against *E. coli* and *P. aeruginosa*.

The antibacterial activity of *Nostoc linckia* is summarized in **Table 18**. In this case, methanolic and ethyl acetate extracts were found active. The methanolic extract exhibited activity against *S. aureus* and *E. coli* as compared to *P. aeruginosa, K. pneumoniae* and *S. typhi* in which no activity was observed. Ethyl acetate extract was found active only against Gram-positive *S. aureus*. Hexane extract, Dichloromethane extract and spent medium did not exhibit any sort of antibacterial activity.

The results for MIC value of *Nostoc linckia* are in **Table 27**. The results showed that the MICs of methanolic extract were >512 μg/ml against *S. aureus* and *E. coli*. The MIC value of ethyl acetate extract was also recorded same as above (*i.e.* >512 μg/ml) against *S. aureus* **(Fig. 40)**.

The results of antibacterial activity of *Nostoc commune* are summarized in **Table 19** showing that methanolic and ethyl acetate extracts were found active. The methanolic extract exhibited activity against *S. aureus* and *E. coli,* whereas no zone of inhibition was observed against *P. aeruginosa, K. pneumoniae* and *S. typhi*. The highest effective zone of inhibition (10.42 mm) was observed against *S. aureus*. Ethyl acetate extract was found active against Gram-positive bacterium *S. aureus* only with the effective zone of inhibition 6.79 mm. Hexane extract, Dichloromethane extract and spent medium did not exhibit any antibacterial activity.

MIC values of the active extracts of *Nostoc commune* are shown in Table-28. The results showed that the MICs of methanolic extracts were more than 512 μg/ml against both *S. aureus* and *E. coli*. The MIC value of ethyl acetate extract was also recorded >512 μg/ml against *S. aureus*.

Solatni *et al.* (2005) demonstrated that the antimicrobial activity of different extracts of *Nostoc* spp. against *Bacillus subtilis, Staphylococcus epidermidis, E. coli, Salmonella typhi* and *Candida albicans* was effective. Similarly, Jaki *et al.* (1999) reported that the antimicrobial activity of *N. muscorum* against Gram-positive bacteria, Gram-negative bacteria and a yeast *Candida albicans,* was significant.

The results of antibacterial activity of *Spirulina platensis* are summarized in Table 20. In this case, all the extracts exhibited activity against *S. aureus*. The methanolic extract showed broad spectrum activity as highest effective zone of inhibition (15.21 mm) was recorded against *S. aureus* followed by *E. coli, P. aeruginosa* and *S. typhi*. However, extracts prepared in hexane, ethyl acetate and dichloromethane were found active only against *S. aureus*. In addition, no inhibitory effect was observed against *K. pneumoniae.* Interestingly, spent medium of this alga was found active against *E. coli*.

The results of minimum inhibitory concentrations of *Spirulina platensis* extracts are shown in **Table 29**. The results showed that the MICs of methanolic extract were 128 µg/ml and 256 µg/ml against *S. aureus* and *E. coli,* respectively. However, the MIC of this extract was recorded >512 µg/ml against *P. aeruginosa* and *S. typhi*. The MIC of dichloromethane extract was found 512 µg/ml against *S. aureus*. The MIC of both the hexane and ethyl acetate extracts were also recorded more than >512 µg/ml.

A number of cyanobacteria species have been found to produce highly anti-HIV active sulfoglycolipids (Reshef *et al.*, 1997; Loya *et al.*, 1998). In addition to their activity, the sulfoglycolipids constitute part of the chloroplast membrane and are, therefore, abundant (Harwood, 1987). Loya *et al.* (1998) concluded that together these facts argue "in favour of further development of these compounds as anti-AIDS agents." Hayashi *et al.* (1996) reported the Anti-HIV activity of extracts of *Spirulina platensis* and they also isolated the compound calcium spirulan (Sulfated polysaccharide).

The present study indicates that the methanolic extracts showed more potent activity against both Gram-positive and Gram-negative bacteria. The similar results had also been reported in determining the antibacterial activity of different extracts and volatile compounds of *Spirulina platensis* (Ozdemir *et al.*, 2004).

Another significant result of the present study was that the spent medium (supernatant) of *Spirulina platensis* showed the activity against *E. coli*. Similarly, Fish and Codd (1994) reported that the extracellular material produced by *Phormedium sp.* inhibited the growth of Gram-positive and Gram-negative bacteria as well as a yeast and a fungus. In another study, Oufdou *et al.* (2001) reported that the extracellular products released by *Pseudoanabaena sp.* reduced the survival of *E. coli, Salmonella* sp., *S. aureus* and *Candida albicans.*

The results of *Westiellopsis prolifica* are summarized in **Table 21**. In this case, all the extracts exhibited activity against a Gram-positive bacterium *S. aureus*. Gram-negative bacteria including *E. coli, P. aeruginosa, K. pneumoniae,* and *S. typhi* were found completely resistant. Methanolic extracts was found more active as compared to hexane, ethyl acetate and dichloromethane. Spent medium was found inactive against all the tested bacteria.

The results of *Hapalosiphon sp.* are presented in **Table 22**, showing that methanolic extract was found active against *S. aureus* and *E. coli*. No zone of inhibition was observed against both *P. aeruginosa* and *S. typhi*. However, *K. pneumoniae* was found resistant when tested with all the extracts. The hexane, ethyl acetate, dichloromethane extracts and spent medium were found inactive against all the test bacteria.

The results of MIC of active extracts of *Westiellopsis prolifica* and *Hapalosiphon sp.* are given in **Table 30** and **31**. The results showed that the MICs of all the extracts were more than 512 μg/ml against all the susceptible bacteria.

Patterson *et al.* (1993) reported the antiviral MIC of lipophilic extracts of *Hapalosiphon intricatus.* Similarly, Rao *et al.* (2007) reported the antimycobacterial activity of different extracts of *Hapalosiphon sp.* and *Westiellopsis prolifica* against five species of *Mycobacterium*. They have also reported the MIC of the active crude extracts. Smitka *et al.* (1992) reported the antifungal activity of *Westiellopsis prolifica*; however, according to our knowledge, no one has yet reported the antibacterial activity in different extracts of this cyanobacterium against *S. aureus*.

The search of literature revealed that the antimicrobial effect of *Nostoc* (Bloor and England, 1991), *Phormidium* (Fish and Codd, 1994), *Anabaena, Oscillatoria, Pseudoanabaena, Synechocystis* (Kreitlow *et al.*, 1999), *Oscillatoria anguistisima* and *Calothrix parietina* (Issa, 1999) extracts on some pathogenic organisms have been reported. They also have reported that the extracts harvested in different solvents were effective against both Gram-positive and Gram-negative organisms. This is in agreement with our findings, since the cyanobacterial species extracted in different solvents had similar effects on both Gram-positive and Gram-negative bacteria used in this study.

In decreasing order, extracts of cyanobacterial species showing considerable antibacterial activity were: *Spirulina platensis, Anabaena variabilis, Nostoc muscorum, N. punctiforme. A. fertilissima,* (each was found active against four bacterial species), *N. linckia* and *N. commune, Hapalosiphon sp.* (each was active against two bacterial species), *Westiellopsis prolifica active* against one bacterial species). The susceptibility of bacterial species to cyanobacterial extracts in decreasing order was: *S. aureus* (susceptible to extracts from all 9 cyanobacterial species), *E. coli* (susceptible to eight cyanobacteria), *P. aeruginosa* (susceptible to five cyanobacteria), and *S. typhi* showing susceptibility to five cyanobacterial species. However, *K. pnumoniae* was found resistant against all the cyanobacterial extracts.

Some studies concerning the effectiveness of extraction methods highlight that methanolic extraction yields higher antimicrobial activity than n-hexane and ethyl acetate (Rosell *et al.*, 1987 and Moreau *et al.*, 1988), whereas others report that chloroform is better than methanol and benzene (Febles *et al.*, 1995). It is clear that using organic solvents always provides a higher efficiency in extracting compounds for antimicrobial activitiy compared to the water-based methods (Asthana *et al.*, 2006; Stensvik *et al.*, 1998). Falch *et al.* (1995) used successive extractions with solvents of increasing polarity *i.e.* petroleum ether, dichloromethane, ethyl acetate, methanol and 80% aqueous methanol. These extracts showed different antibacterial effects in bioautographic assay with *B. subtilis*, *E. coli* and *Micrococcus luteus*. According to our experimental results, methanol caused better halo-zones than ethyl acetate, hexane and dichloromethane against both Gram-positive and Gram-negative bacteria used in the study.

In comparison with the standard ciprofloxacin (fig. 36), all the extracts exhibited low antibacterial activity. Maximum antibacterial activity was shown against *S. aureus* followed by *E. coli*, *P. aeruginosa* and *S. typhi*. *Klebsiella pneumoniae* was found to be least susceptible against all tested cyanobacterial extracts. No any zone of inhibition was observed when pure solvents were analyzed for antibacterial assay **(Fig. 37)**. The results obtained from the present study indicate development of antimicrobial substances from different cyanobacterial extracts and their promising utilization in treatment of various infectious diseases.

Chapter 7

SUMMARY

Present work was aimed to determine the antibacterial activity and minimum inhibitory concentration (MIC) of active crude extracts of selected cyanobacterial species. Nine cyanobacterial strains *viz. Anabaena variabilis, A. fertilissima, Nostoc muscorum, N. punctiforme, N. linckia, N. commune, Spirulina platensis, Westiellopsis prolifica* and *Hapalosiphon sp.* belonging to family; Nostocaceae, Oscillatoriaceae and Stigonemataceae were selected from the herbarium of Indian Agricultural Research Institute, New Delhi, and maintained in BG-11 growth medium throughout the study in the laboratory.

Five pathogenic bacteria *viz., Staphylococcus aureus, Escherichia coli, Pseudomonas aeruginosa, Klebsiella pneumoniae* and *Salmonella typhi* were collected from Institute of Microbial Technology (IMTECH), Chandigarh, and maintained throughout the study on nutrient agar slants with sub-culturing on fresh slant. All the media used in the study were procured from Hi-media and chemicals were of analytical grade.

The growth curve for each culture was studied by growing them in Erlenmeyer flask using BG-11 growth medium. Optical Densities (OD) were recorded at 650 nm by using UV-Vis spectrophotometer and then were plotted against number of incubation days for each cyanobacterium. They were examined to determine the early stationary phases of their life-cycle, which are believed to be the most appropriate periods for the production of secondary metabolites by an organism.

After 23-25 days of incubation period, each culture was harvested and spent medium was collected by filtration of the biomass. Hexane, ethyl acetate, dichloromethane and methanol were used as the solvents for obtaining the extracts from biomass of each culture. The extracts and spent medium thus obtained were then investigated for antibacterial activity against bacterial strains. The extracts found active in antibacterial assay were subjected for the determination of minimum inhibitory concentration against bacterial species found susceptible in antibacterial assay.

Antibacterial activity of all the extracts was examined by agar well diffusion assay. Minimum inhibitory concentration of active crude extracts was determined by 96-well micro broth dilution method as recommended by National Committee for Clinical Laboratory Standards.

Antibacterial activity of cyanobacterial species was observed against both Gram-positive and Gram-negative bacteria. In antibacterial assay, the methanol extract showed more potent activity than hexane, dichloromethane, ethyl acetate and spent medium. The antibacterial activity of extracts varied with the test organisms. Compratively lower MIC value was obtained against Gram-positive bacteria.

Methanolic extract of *Anabaena variabilis* exhibited greater effective zone of inhibition against *Staphylococcus aureus,* whereas no zone of inhibition was observed against *Klebsiella pneumoniae*. Hexane extracts showed little activity against *S. aureus*. Ethyl acetate extract was found active against *S. aureus* and *Escherichia coli* only. Dichloromethane extract and spent medium did not exhibit any antibacterial activity. Furthermore, the active extracts obtained from the above process *i.e.* hexane, ethyl acetate, and methanolic extracts were then subjected to the determination of minimum inhibitory concentrations by 96-well micro broth dilution method. Active crude extracts were diluted in CAMHB in the range from 512 μg/ml to 1 μg/ml using ciprofloxacin as a standard antibiotic. Methanol extract of *A. variabilis* showed best MIC values against *E. coli* and S. *aureus,* it was more than 512 μg/ml when tested against *P. aeruginosa* and *S. typhi*. The MIC of ethyl acetate extract was observed as 512 μg/ml against *S. aureus* but it was more than 512 μg/ml against *E. coli*. The MIC of hexane extract was found more than 512 μg/ml when tested against *S. aureus*.

In the case *Anabaena fertilissima* only two extracts *i.e.* methanolic and ethyl acetate extracts were found active as compared to previously studied *Anabaena variabilis*. The highest effective zone of inhibition was observed in methanolic extract against *S. aureus* followed by *E. coli, P. aeruginosa* and *S. typhi*. Hexane extract, dichloromethane extract and spent medium were found inactive against all examined bacteria. *K. pneumoniae* was found completely resistant to all experimented extracts. Methanolic extract having the great MIC values 512 µg/ml against *S. aureus* and *E. coli*, however, it was more than 512 µg/ml when tested against *P. aeruginosa* and *S. typhi*. The MIC of ethyl acetate extract was found > 512 µg/ml against *S. aureus* and *E. coli*.

Methanolic extract of *Nostoc muscorum* was found active against all the tested microorganisms except *K. pneumoniae*. The highest effective zone of inhibition was observed against *S. aureus* followed by *E. coli*, P. *aeruginosa* and *S. typhi*. Hexane extracts showed little activity against *S. aureus*. Ethyl acetate extract was found active against *S. aureus* and *E. coli*. Dichloromethane extract and spent medium were found inactive against all the test bacteria. The results showed that the MIC of methanolic extract were found greater against *S. aureus* and *E. coli* whereas it was >512 µg/ml when tested against *P. aeruginosa* and *S. typhi*. The MIC of ethyl acetate extracts was 512 µg/ml against S. *aureus* and *E. coli*. The MIC of hexane extract was also found to be >512 µg/ml against *S. aureus*.

In the case of *Nostoc punctiforme*, only methanolic and ethyl acetate extracts were found active. The highest effective zone of inhibition was observed in methanolic extract against *S. aureus* followed by *E. coli, P. aeruginosa* and *S. typhi*. Hexane extract, dichloromethane extract and spent medium were found inactive against both types of bacteria. No activity was observed against *K. pneumoniae*. The minimum inhibitory concentrations of the active extracts of *N. punctiforme* showed that the MICs of methanolic extract was recorded 512 µg/ml against both *E. coli* and *S. aureus* whereas it was more than 512 ìµg/ml against *P. aeruginosa* and *S. typhi*. The MIC of ethyl acetate extract was 512 µg/ml against *S. aureus* but it was recorded >512 µg/ml against *E. coli* and *P. aeruginosa*.

The methanolic extract of *Nostoc linckia* exhibited activity against *S. aureus* and *E. coli* as compared to *P. aeruginosa, K. pneumoniae* and *S. typhi* in which no activity was observed. Ethyl acetate extract was found active only against Gram-positive bacterium *S. aureus*. Hexane extract, dichloromethane extract and spent medium did not exhibit any sort of antibacterial activity. The MICs of all the active extracts was recorded > 512 µg/ml.

Only methanolic and ethyl acetate extracts of *N. commune* were found active. The methanolic extract exhibited activity against *S. aureus* and *E. coli,* whereas no zone of inhibition was observed against *P. aeruginosa, K. pneumoniae* and *S. typhi*. The highest effective zone of inhibition was observed against *S. aureus.* Ethyl acetate extract was found active against Gram-positive bacterium *S. aureus* only. Hexane extract, Dichloromethane extract and spent medium did not exhibit any antibacterial activity. The MICs of all the active extracts was recorded > 512 µg/ml.

All the extracts of *Spirulina platensis* exhibited activity against *S. aureus*. The methanolic extract showed broad spectrum activity as highest effective zone of inhibition was recorded against *S. aureus* followed by *E. coli, P. aeruginosa* and *S. typhi*. However, extracts prepared in hexane, ethyl acetate and dichloromethane were found to be active only against *S. aureus*. In addition, no inhibitory effect was observed against *K. pneumoniae.* Interestingly, spent medium of this alga was found active against *E. coli*. Methanolic extracts of this alga showed lower MIC value against *S. aureus* and *E. coli* however it was found to be more than 512 µg/ml against *P. aeruginosa* and *S. typhi*. The MIC of dichloromethane extract was found to be 512 µg/ml against *S. aureus.* The MIC of both the hexane and ethyl acetate extracts was recorded more than 512 µg/ml.

In *Westiellopsis prolifica,* all the extracts exhibited activity against a Gram-positive bacterium *S. aureus* only. Gram-negative bacteria including *E. coli, P. aeruginosa, K. pneumoniae,* and *S. typhi* were found completely resistant. Methanolic extract was found more active as compared to hexane, ethyl acetate and dichloromethane. Spent medium was found inactive against all the tested bacteria. The MICs of all the active extracts was recorded > 512 µg/ml.

Methanolic extract of *Hapalosiphon sp.* was found active against *S. aureus* and *E. coli*. No zone of inhibition was observed against both *P. aeruginosa* and *S. typhi*. However, *K. pneumoniae* was found resistant when tested with all the extracts. The hexane, ethyl acetate, dichloromethane extract and spent medium were found inactive against all the test bacteria. The MICs of all the active extracts was recorded > 512 µg/ml.

The above findings indicate the discovery of novel chemicals that can lead to the development of pharmaceuticals by cyanobacteria. The current scenario of antibiotics is very threatening with significant emergence of resistance among bacterial pathogens against available antibiotics. The present investigation reveals that cyanobacteria would be the major source in finding such metabolites with greater efficacy.

Further study can be made to isolate the pure compounds responsible for the activity from the extracts with the help of column chromatography. The isolated compounds can be further identified by various techniques like GC-MS, LC-MS, IR and NMR.

Chapter 8

CONCLUSION

The objective of the present investigation was to evaluate the antibacterial properties of cyanobacterial species *viz., Anabaena variabilis, Anabaena fertilissima, Nostoc muscorum, Nostoc punctiforme, Nostoc linckia, Nostoc commune, Spirulina platensis., Westiellopsis prolifica* and *Hapalosiphon sp.* against some pathogenic species including *Staphylococcus aureus, Escherichia coli, Pseudomonas aeruginosa, Klebsiella pnumoniae* and *Salmonella typhi.* The following are the major findings of the study.

- Extracts prepared in methanol were found more active as compared to ethyl acetate, hexane and dichloromethane.
- Antibacterial activity of extracts varied with the test organisms. Maximum antibacterial activity was shown against *Staphylococcus aureus* followed by *Escherichia coli, Pseudomonas aeruginosa* and *Salmonella typhi. Klebsiella pneumoniae* was found to be least susceptible against all tested cyanobacterial extracts.
- Antibacterial activity was observed against both Gram-positive and a spectrum of Gram-negative bacteria by almost all the cyanobacterial species examined in the present study except *Westiellopsis prolifica* where activity was observed only against *S. aureus,* a Gram-positive bacterium.

- Methanolic extract of *Anabaena variabilis* exhibited greater effective zone of inhibition against *Staphylococcus aureus* followed by *Escherichia coli, Pseudomonas aeruginosa* and *Salmonella typhi* However, dichloromethane extract was found inactive.
- Only ethyl acetate and methanolic extract of *Anabaena fertilissima* exhibited effective zone of inhibition.
- Methanolic extract of *Nostoc muscorum* exhibited greater effective zone of inhibition against *S.aureus* followed by *E. coli, P. aeruginosa* and *S. typhi*.
- Only ethyl acetate and methanolic extract of *Nostoc punctiforme* exhibited effective zone of inhibition.
- Methanolic extract of *Nostoc linckia* and *Nostoc commune* exhibited activity against *S.aureus* and *E.coli*. Little activity of ethyl acetate extracts was found against *S.aureus*.
- All the organic extracts of *Spirulina platensis* were found active, with the greatest effective zone of inhibition observed against *S.aureus*.
- Extracellular substances (Spent medium) of *Spirulina platensis* was found active only against *E.coli*.
- Only *S. aureus* was found susceptible against all the organic extracts of *Westiellopsis prolifica*.
- Methanolic extract of *Hapalosiphon sp.* was found active only against *S.aureus* and *E.coli*.
- Lower Minimum inhibitory concentration values were observed against *S. aureus*.
- In comparison with the standard antibiotic ciprofloxacin, all the extracts exhibited low antibacterial activity.

The above findings indicate the discovery of novel chemicals that can lead to the development of pharmaceuticals by cyanobacteria. The current scenario of antibiotics is very threatening with significant emergence of resistance among bacterial pathogens against available antibiotics. The present investigation reveals that cyanobacteria would be a major source in finding such metabolites with greater efficacy.

Chapter 9

LITERATURE CITED

Accorinti, J.1983. Antifungal products from algal origin. *Rev Int Ocean Med*, 72, 45-53.

Allnutt, F.C.T. 1996. Cyanobacteria (Blue -Green Algae) Biotechnology: Past, Present and Future. *Journal of Scientific and Industrial Research*. 55, 693-714.

Ananthanarayan, R. and Paniker, C.K.J. 2005. Textbook of Microbiology (VII Ed.) Orient Longman Pvt Ltd, Chennai-600 002. pp 1-657.

Asthana, R.K., Srivastava, A., Singh, A.P., Deepali, Singh, S.P., Nath, G., Srivastava, R. and Srivastava, B.S. 2006. Identification of an antimicrobial entity from the cyanobacterium *Fischerella sp.* isolated from bark of *Azadirachta indica* (Neem) tree. *Journal of Applied Phycology*. 18, Number-1, 33-39.

Ayehunie, S., Belay, A., Baba, T.W., and Ruprecht, R.M. 1998. Inhibition of HIV-1 replication by an aqueous extract of *Spirulina platensis* (*Arthrospira platensis*). *J. Acquir. Immune. Defic. Syndr. Hum. Retrovirol.* 18, 7-12.

Bagasra, O., Lischner, H.W., Halligen, G.E., and Heins, B. 1989. Anti-HiV-1 activities of sulfated monosaccharides and polysaccharides: Therapeutic indices. Int. Conf. AIDS 5, 660.

Ballantine, D.L., Gerwick, W.H., Velez, S.M., Alexander, E. and Guevara, P. 1987. Antibiotic activity of lipid-soluble extracts from Caribbear marine algae. *Hydrobiologia*, 151/152, 463-469.

Ballesteros, E., Martin, D., and Uriz, M.J. 1992. Biological activity of extracts from some Mediterranean macrophytes. *Bot Marina* 35, 481-485.

Banker, R. and Carmeli, S. 1998. Tenuecyclamides A-D, cyclic hexapeptides from cyanobacterium *Nostoc spongiaeforme* var.*tenue*. *J.Nat.Prod.* 61, 1248-1251.

Barchi, J.J.Jr., Moore,. R.E. and Patterson, G M.L.1984. Acutiphycinn and 20, 21-didehydroacutiphycin, new antineoplastic agents from the cyanophyte *Oscillatoria acutissima*. *J. Am. Chem. Soc.* 106, 8193-8197.

Beress, A., Wassermann, O., Tahhan, S., Bruhn, T., Beress, L., Kraiselburd, E.N., Gonzalez, L. V., de Motta, G.E., and Chavez, P.I. 1993. A new procedure for the isolation of anti-HIV compounds(polysaccharides and polyphenols) from the marine alga *Fucus vesiculosus*. *J. Nat. Prod.* 56, 478-488.

Bernart, M.W., Whatley, G.G., and Gerwick, W.H. 1993. Unprecedented

Bladon, C.M. 2002. Pharmaceutical Chemistry. Wiley, UK.

Blinkova, L.P., Gorobets, O.B. and Baturo, A.P. 2001. Biological activity of *Spirulina*. *Zh mikrobiol Epidemiol Immunobiol.* 2, 114-8.

Bloor, S. and England, R.R. 1991. Elucidation and optimization of the medium constituents controlling antibiotic production by the cyanobacterium *Nostoc muscorum*. *Enzym Microbiol. Technol.* 13, 76-81.

Bloor, S. and England, R.R.1989. Antibiotic production by the cyanobacterium *Nostoc muscorum*. *Journal of Applied Phycology*. 1, 367-372.

Bonjouklian, R., Smitka, T.A., Doolin, L.E., Molloy, R.M., Debono, M., Shaffer, S.A., Moore, R.E., Stewart, J.B. and Patterson, G.M.L. 1991 Tjipanazoles, new antifungal agent from the blue green alga *Tolypothrix tjipanasensis*. *Tetrahedron.* 47, 7739-7750.

Bonjouklian, R., Moore, R.E., Mynderse, J.S., Patterson, G.M.L. and Smitka, T.A.1988. New hapalindoles. European patent application EP 0260964 A2, 1-26.

Boyd, M.R., Gustafson, K., McMahon, J., and Shoemaker, R.,1996. Discovery of cyanovirin-N, a novel HIV-inactivating protein from *Nostoc ellipsosporum* that targets viral gp120. Int. Conf. AIDS 11, 71-71.

Boyd, M.R., Gustafson, K.R., McMahon, J.B., Shoemaker, R.H., Okeefe, B.R., Mori, T., Gulakowski, R.J., Wu. L., Rivera, M.I., Laurencot, C.M., Currens, M.J., Cardellina, I.I., Henderson, L.E. 1997. Discovery of cyanovirin-N, a novel human immunodeficiency virus-inactivating protein that binds viral surface envelope glycoprotein gp120: Potential application to microbiocide development. *Antimicrob. Agent Chemother*. 41, 1521-1530

Bradner, W.T., 1980. Fermentation approaches to new drug development. In Crooke, S.T. & Prestayko, A.W. *Cancer and Chemotherapy*. Academic Press, New York. Pp. 313-25.

Browitzka, M.A., 1995. Microalgae as a source of pharmaceuticals and other biologically active compounds. *J. Appl. Phycol*. 7, 3-15.

Brown, R. W. 1956. Composition of Scientific Words. Smithsonian Institutional Press. ISBN 0-87474-286-2.

Burja, A.M., Banaigs, E.B., Abou-Mansour, B.J.G. and Wright, P.C. 2001. Marine cyanobacteria - a prolific source of natural products. *Tetrahedron*. 57, 9347-9377.

Burkholder, P.R., Burkholder, L.M. and Almodovar, L.R. 1960. Antibiotic activity of some marine algae of Puerto rico. *Bot Marina*. 2, 149-156.

Cannell, R.J.P., Owsianka, A.M. and Walker, J.M. 1988. Results of a Large-Scale Screening Programme to Detect Antibacterial Activity from Freshwater Algae. *Br. Phycol. J.* 23, 41-44.

Cardellina, J.H., Moore, R.E., Arnold, E.V. and Clardy, J. 1979. Structure and absolute configuration of malyngolide, an antibiotic from the marine blue green alga *Lyngbya majuscula Gomont. J. Org. Chem.* 44, No-23, 4039-4042.

Carmeli, S., Moore, R.E., and Patterson, G.M.L., Corbett, T.H., and Valeriote, F.A. 1990 c.Tantazoles, unusual cytotoxic alkaloids from the blue-green alga *Scytonema mirabile. J. Am. Chem. Soc.* 112, 8195-8197.

Carmeli, S., Moore, R.E.and Patterson, G.M.L. 1990 a. Isonitriles from the blue-green alga *Scytonema mirabile. J. Org.. Chem.* 4431-8.

Carmeli, S., Moore, R.E.and Patterson, G.M.L. 1990 b. Tolytoxin and New Scytophycins from three species of *Scytonema. J. Nat. Prod.* 53, No-6, 1533-1542.

Carmichael, W.W. 1992. Cyanobacterial secondary metabolites-the cyanotoxin. *J. Appl. Bacteriol.* 72, 445-459.

Centeno, P.O.R., Ballantine, D.L. and Gerwick, W.H. 1996. Dynamics of antibacterial activity in three species of Caribbean marine algae as a function of habitat and life history. *Hydrobiologia,* 326/327, 457-462.

Chang, T. 1993. Antibiotic substances produced by a marine green alga, *Dunaliella primolecta, Bioresource Technology*. 44:2, 149-153.

Chen, J.L., Moghaddam, M.F.and Gerwick, W.H. 1993. Gloiosiphones A and B, novel metabolite from the red marine alga *Gloiosiphonia verticillaris. J. Nat. Prod.* 56, 1205-1210.

Chetsumon, A. 1995. Continouous antibiotic production by an immobilized cyanobacterium in a seaweed type bioreactor, *Journal of Applied Phycology*. 7:2, 135-139.

Chorus, I. and Bartram, J. 1999. Toxic cyanobacteria in the water. E & F.N. Spon, London.

Choudhury, S., Sree, A., Mukherjee, S.C., Pattnaik, P. and Bapuji, M. 2005. *In vitro* antibacterial activity of extracts of selected marine algae and mangroves against fish pathogens. Asian Fisheries Science. 18, 285-294.

Chrost, R.J. 1975. Inhibitors produced by algae as an ecological factor affecting bacteria in water ecosystem. Acta Microbiol Pol Ser B, 125-133.

Damonte, E.B., Neyts, J., Pujol, C.A., Snoeck, R., Andrei, G., Ikeda, S., Witvrouw, M., Reymen, D., Haines, H., and Matulewicz, M.C. 1994. Antiviral activity of a sulfated polysaccharide from the red seaweed *Nothogenia fastigiata. Biochem. Pharmacol.* 47, 2187-2192.

De Cano, M.M.S., De Mule, M.Cr., De Caire, G.Z., and De Halperin, D.R. 1990. Inhibition of *Candida albicans* and *Staphylococus aureus* by phenolic compounds from terrestrial cyanobacterium *Nostoc muscorum. Journal of Applied Phycology*. 2, No-1, 79-81.

De PK, Sulaiman M (1950) The influence of algal growth in the rice fields on the yield of crops. Indian J Agric Sci 20:327-342.

Debro, L.H. and Ward, H.B. 1979. Antibacterial Activity of Freshwater Green Algae. *Planta Medica*. 36, 375-378.

Desikachary, T.V. 1959. Cyanophyta. Indian Concil of Agricultural Research, New Delhi, India, pp. 1-686.

Dhananjaya, P. S., Kumar, A., Tvagi, M. B.2003.Biotoxic cyanobacterial metabolites exhibiting pesticidal and mosquito larvicidal activities. *Journal of microbiology and biotechnology*. vol. 13, no1, pp. 50-56.

Doan, N.T., Rickards, R.W., Rothschild, J.M. and Smith,G.D. 2000. Allelopathic action of the alkaloid 12-epi-hapalindole E isonitrile and calothrixin A from cyanobacteria of the genera *Fischerella* and *Calothrix*. *Journal of Applied Phycology*. 12, 409-416.

Donia, M., and H. T. 2003. Mrine natural products and their potential application as anti-infective agents. *The Lancet*. 3, 338-348.

Duy, T.N., Lam, P.K., Shaw, G.R. and Connell, D.W. 2000. Toxicological and risk assessment of freshwater cyanobacterial (blue-green algae) toxin in water. *Rev. Environ. Contam. Toxicol*. 163, 113-185.

Espeche, M.E., Fraile, E.R.and Mayer, A.M.S. 1984. Screening of Argentine marine algae for antimicrobial activity. *Hydrobiologia*, 116/117, 525-528.

Falch, B.S. 1996. Was stacks in cyanobakterien? Pharmazie in unserer Zeit 25, 311-321.

Falch, B.S., Konig, G.M., Wright, A.D. 1992. Antibacterial and cytotoxic compounds from blue-green algae *Fischerella ambigua. Planta Medica*, 58, 654-655.

Falch, B.S., Konig, G.M., Wright, A.D. and Sticher, O. 1993. Ambigol A and B: New Biologically Active polychlorinated Aromatic Compounds from the Terrestrial Blue-Green Alga *Fischerella ambigua. J. Org. chem.* 58, 6570-6575.

Falch, B.S., Konig, G.M., Wright, A.D., Sticher, Angerhofer, C.K., Pezzuto, J.M. and Bachmann, H. 1995. Biological Activities of Cyanobacteria: Evaluation of Extracts and Pure Compounds. *Planta Med.* 61, 321-328.

Fatma, T. and Venkataramall, V. 1994. Cyanobacteria and Microalgae Potential as biochemicals, *Journal of Timber Forest Products*.1, 19-39.

Febles, C.L., Arias, A., Gill-Rodriguez MC *et al.* 1995. *In vitro* study of antibacterial activity in algae (Chlorophyta, Phaeophyta and Rhodophyta) collected from the coast of Tenerife (in Spanish). Anuario del Estudios Canarios. 34, 181-192.

Feng, P., Weagant, S., Grant, M. 2002. Enumeration of *Escherichia coli* and the Coliform Bacteria. Bacteriological Analytical Manual (8th ed.). FDA/Center for Food Safety & Applied Nutrition.

Filho, J.V.M.L., Carvalho, A.F.F.U., Freitas, S.M. and Melo, M.M. 2002. Antibacterial Activity of extracts of six microalgae from the northeastern Brazilian coast. *Brazillian Journal of Microbiology*. 33, 311-313.

Fish, S.A. and Codd, G.A. 1994. Bioactive compound production by thermophilic and thermotolerant cyanobacteria (blue-green algae). *World J. Microbiol. Biotech.* 10, 338-341.

Flores, E. and Wolk, C.P. 1986. Production, by filamentous, nitrogen- fixing cyanobacteria, of a bacteriocin and other antibiotics that kill related strains. *Arch Microbiol.* 14, 215-219.

Fotadar, U., Zaveloff, P., Terracio, L. 2005. "Growth of *Escherichia coli* at elevated temperatures". *J. Basic Microbiol.* 45, 5 403–4.

Frankmolle, W .P., Larsen, L.K., Caplan, F.R., Patterson, G.M.L., Knubel, G, Levine, I.A., and Moore, R.E., 1992. Antifungal cyclic peptides from the terrestrial blue green alga *Anabaena laxa*. Isolation and biological properties. *J antibiotics*. 45, 1451-1457.

Fujii, K., Sivonen, K., Adachi, K., Noguchi, K., Sano, H., Hirayama, K., Suzuki, M. and Harada, K. 1997. Comprative study of toxic and non-toxic cyanobacterial products: Novel peptides from toxic *Nodularia spumigena* A V1. *Tetrahedron Lett.* 38, 5525-5528.

Gerber, P., Dutcher, J.D., Adams, E.V., and Sherman, J.H. 1958. Protective effect of seaweed extracts for chicken embroyes infected with influenza B or mumps. *Proc. Soc. Exp. Bio. Med.* 99, 590-593.

Gerwick, W.H., Roberts, M.A., Proteau, P.J. and Chen, J.L. 1994. Screening cultured marine microalgae for anticancer-type activity. *J Appl. Phycol.* 6, 143-149

Gerwick, W.H.M., Mrozek, C., Moghaddam, M.F., and Agarwal, S.K. 1989. Novel cytotoxic peptides from the tropical marine cyanobacterium *Hormothamnion entermorphoides*. Discovery, isolation and initial chemical and biological characterization of the hormothamnins from wild and cultured material. *Experientia.* 45, 115-121.

Ghasemi, Y., Yazdi, M.T., Shafiee, A., Amini, M., Shokravi, S. and Zarrini, G. 2004. Parsiguine, A Novel Antimicrobial substance from *Fischerella ambigua. Pharmceutical Biology.* 42, No-5, 318-322.

Giannella, R. A. 1996. *Salmonella.* In: Baron's Medical Microbiology , 4th ed., Univ of Texas Medical Branch. (via NCBI Bookshelf) ISBN 0-9631172-1-1.

Glombitza, K.W. and Koch, M. 1989. Secondary metabolites of pharmaceutical potential. In Algal and Cyanobacterial Biotechnology ed. Cresswell, R.C., Rees, T.A.V. and Shah, N.pp. 161-238. Harlow, UK: Longman Scientific & Technical.

Gol'din, E.B., 2003. Antibacterial activity of pure culture of cyanobacteria and algae. *Mikrobiol.* Z. 65, (4) 68-76.

Goyal, P., Khanna, A., Chauhan, A., Garima and Kaushik, P. *In Vitro* Evaluation of Crude Extracts of *Catharanthus roseus* for Potential Antibacterial Activity accepted in International Journal of Green Pharmacy.

Gupta, A.B. and Shrivastava, G.C. 1965. On Antibiotic properties of some fresh water algae. *Hydrobiologica.* 25, 285-288.

Gustafson, K. R., Sowder, R., Henderson, L., Pannel, L., Cardellina, J., McMahon, J., Shoemaker, R., and Boyd, M.,1996. Sequence determination and chemical chractristics of a novel anti-HIV protein, Cyanovirin-N, isolated from the cyanobacterium *Nostoc ellipsosporum.* Int. Cont. AIDS, 11, 70.

Gustafson, K.R., Cardellina II, J.H., Fuller, R.W., Weislow, O.S., Kiser, R.F., Sander, K.M., Patterson, G.M.L., Boyd, M.R., 1989. AIDS-antiviral sulfolipids from cyanobacteria (blue-green algae). *J. nat. Cancer Inst.* 81, 1254-1258.

Hagmann, L. and Juttner, F. 1996. Fischerellin A, a novel photosystem-II-inhibiting alleochemical of the cyanobacterium *Fischerella muscicola* with antifungal and herbicidal activity. *Tetrahedron Lett.* 37, 6539-6542.

Harrigan, G.G., Luesch, H., Yoshida, W.Y., Moore, R.E., Nagle, D.G., Paul, V.J., Mooberry, S.L., Corbett, T.H., Valeriote, F.A., 1998b. Isolation, Structure Determination, and biological activity of dolastatin 12 and lyngbyastatin 1 from *Lyngbya majuscule/ Schizothrix calcicola* cyanobacterial assemblages. *J. Nat. Prod.* 61, 1221-1225.

Harrigan, G.G., Luesch, H., Yoshidi, W.Y., Moore, R.E., Nagle, D.G., Paul, V.J., 1999. Symplostatin 2; A dolstatin 13 analogue from the marine cyanobacterium *Symploca hydnoides. J. Nat. Prod.* 62, 655-658.

Harrigan, G.G., Luesch, H., Yoshidi, W.Y., Moore, R.E., Nagle, D.G., Paul, V.J., Mooberry, S.L.Corbett, T.H., Valeriote, F.A., 1998a. Symplostatin 1: A dolastin 10 analogue from the marine cyanobacterium *Symploca hydnoides. J. Nat. Prod.* 61, 1075-1077.

Harwood, J. 1987. Sulfolipids. In P.K. Stumpf (Ed.), vol.9, pp. 275-289.The Biochemistry of plants. Academic Press, New York.

Hasui, M., Matsuda, M., Okutani, K., and Shigeta, S., 1995. *In vitro* antiviral activity of sulfated polysaccharides from marine microalga (*Cochlodinium polykrikoides*) against HIV and other enveloped viruses. *Int. J. Biol. Macromol.* 17, 293-297.

Hayashi, K., Hayashi, and Morita, M. 1992. An extract of *Spirulina platensis* is a selective inhibitor of herpes simplex virus type-1 penetrationin to HeLa cells. *Phytother. Res.* 7, 7680-7680.

Hayashi, K., Hayashi, T., Kojima, J. 1996. A natural sulfated polysaccharide, calcium spirulan, isolated from *Spirulina platensis*: *in vitro* and *ex vivo* evaluation of anti-herpes simplex virus and anti-human immunodeficiency virus activities. AIDS *Res. Hum. Retroviruses.* 12, 1463-1471.

Hayashi, O., Hayashi, T., Katoh, T., Miyajima, H., Hirano, T., and Okuwaki, Y. 1998. Class-specific influence of dietrary *Spirulina platensis* on antibody production in mice. *J. Nutr. Si. Vitaminol.* 44, 841-851.

Hayashi, O., Katoh, T. and Okuwaki, Y. 1994. Enhancement of antibody production in mice by dietary *Spirulina platensis. J. Nutr. Sci. Vitaminol* (Tokyo).

Hayashi, T., Hayashi, K., Maeda, M. and Kojima, I. 1996. Calcium spirulan, An inhibitor of enveloped virus replication, from a blue- green alga *Spirulina platensis. J. Nat. Prod.* 59, 83-87.

Henning, M., and Kohl, J.G. 1981. Toxic blue- green algae blooms found in some lakes in the German Democratic Republic. *Int. Rev. Ges. Hydrobiol.* 66, 553-561.

Herrero A and Flores E ., 2008. The Cyanobacteria: Molecular Biology, Genomics and Evolution, 1st ed., Caister Academic Press. ISBN 978-1-904455-15-8 .

Heyduck, S.B. and Fischer, U. 2000. Extracellular substances inhibit microbial growth. *Int Microbiol.* 3 (4), 231-4.

Heyman, D. 2004. Control of Communicable Diseases Manual.18th

Edition. Washington DC: American Public Health Assocation.

Holt, J.G., Krieg, N.R., Sneath, P.H.A., Staley, J.T. and Williams, S.T. 1994. Bergey's Manual of Determinative Bacteriology (9th edition). Williams & Wilkins, 428, East Perston Street, Baltimore, Maryland 21202, USA. pp.VII+787.

Horikawa, M., Noro, T. and kamei, Y. 1996. Characterization and purification of antibacterial substances from a red algae, *Laurencia okamurae,* Marine and Highland Bioscience Centre Report, Japan 3, 45-42.

Hoshino, T., Hayashi, T., Hayashi, K., Hamada, J., Lee, J.B., and Sankawa, U. 1998. An antivirally active sulfated polysaccharide from *Sargassum horneri*. *Biol. Pharm. Bull.* 21, 730-734.

Hunter, P.R. 1995. Cyanobacterial toxins and their potential risk to drinking water supplies. *Microbiol. Eur.* 3, 8-10.

Ikawa, M., Haney, J.F., Sasner, J.J. 1996. Inhibition of *Chlorella* growth by the lipids of cyanobacterium *Microcystis aeruginosa*. *Hydrobiol.* 331, 167-170.

Immunosuppressive peptide from the blue-green alga *Lyngbya*

Inderjit and Dakshini, K.M.M. 1994. Algal allelopathy. *The Botanical Review.* 60, 182-196.

Indian Pharmacopoeia. 2007. Microbiological assay of antibiotics (2.2.10). The Indian Pharmacopoeia Commission, Central Indian Pharmacopoeia Laboratory, Government of India, Ministry of Family Welfare, Ghaziabad, 201002.

Ishibashi, M., Moore, R.E., Patterson, G.M.L., Xu C. and Clardy,J., 1986. Scytophycins, cytotoxic and antimycotic agents from the Cyanophyte *Scytonema Pseudohofmanni*. *J. Org. Chem.* 51, 5300-5306.

Ishida, K., Matsuda, H., Murakami, M., Yamaguchi, K. 1997a. Micropepiins 478-A and-B, plasmid inhibitors from the cyanobacterium *Microcystis aeruginosa*. *J. Nat. Prod.* 60, 184-187.

Ishida, K., Matsuda, H., Murakami, M., Yamaguchi, K. 1997b. Microginins 299-A and-B, leucine aminopeptidase inhibitors from the cyanobacterium *Microcystis aeruginosa* (NIES). *Tetrahedron*. 53, 10281-10288.

Ishida, K., Matsuda, H., Murakami, M., Yamaguchi, K. 1997c. Kawaguchipeptin B, an antibacterial cyclic undecapeptide from the cyanobacterium *Microcystis aeruginosa*. *J. Nat. Prod.* 60, 724-726.

Isao, S., Kazutake, Takada, Yasuyuki Sakai and Tetsu Tatsuma. 2005. Compact amperometric algal biosensors for the evaluation of water toxicity. *Analytica Chimica Acta,* Volume 530, Issue 2, 14 , Pages 191-197

Issa, A.A. 1999. Antibiotic production by the Cyanobacteria *Oscillatoria angustissima* and *Calothrix parietina*. *Env. Toxic. Pharm.* 8, 33-37.

Jaki, B., Heilmann, J. and Sticher, O. 2000. New Antibacterial Metabolites from the Cyanobacterium *Nostoc commune* (EAWAG 122b). *J. Nat. Prod.* 63, 1283-1285.

Jaki, B., Orjala, J., Burgi, H.R. and Sticher, O. 1999. Biological Screening of Cyanobacteria for Antimicrobial and Molluscicidal Activity, Brine Shrimp Lethality, and Cytotoxicity. *Pharmaceutical Biology.* 37, No. 2. 138-143.

Jaki, B., Orjala, J., Sticher, O. 1999. A novel extracellular diterpenoid with antibacterial activity from the cyanobacterium *Nostoc commune*. *J. Nat. Prod.* 62, 502-503.

Jaki, B., Zerbe, O., Heilmann, J. and Sticher, O. 2001. Two Novel Cyclic Peptides with Antifungal Activity from the Cyanobacterium *Tolypothrix byssoidea (EAWAG 195)*. *J. Nat. Prod.* 64, 154-158.

Jaspars, M. and Lawton, L.A. 1998. Cyanobacteria-a novel source of pharmaceuticals. *Current Opinion Drug Discovery and Development* 1 (1), 77-84.

Jensen, G.S., Ginsberg, D. I., Huerta, P., Citton, M., and Drapeau, C. 2000. Consumption of Aphanizomenon flos aquae has rapid effects on the circulation and function of immune cells in humans. A novel approach to nutritional mobilization of the immune system.JANA. 2, 50-58.

Jones, G.J. and Korth, W. 1995. *In situ* production of volatile odor compounds by river and reservoir phytoplankton population in Australia. *Water Sci. Technol.* 31, 145-151.

Kajiyama, S., Kanzaki, H., Kawazu, K. and Kobayashi, A. 1998. Nostofungicidine, an Antifungal Lipopeptide from the Field-grown Terrestrial Blue-Green Alga *Nostoc commune*. *Tetrahedron Letters*. 39, 3737-3740.

Katircioglu, H., Beyatli, Y., Aslim, B., Yuksekdag, Z. and Atici, T. 2006. Screening of antimicrobial Agent Production of Some Microalgae in Freshwater. *The Internet Journal of Microbiology*. 2, No.-2, 1-9.

Katiyar, S. K. and Katiyar, R. 2009. Medicinal and Antibacterial properties of Livestock of Lansdowne (Garhwal). In P. Kaushik (Ed.) Indigenous Medicinal Plants Including Microbes and Fungi. Today and Tomorrow's Printers and Publishers, New Delhi.

Kaushik, B. D. 1994. Blue green algae and sustainable agriculture. In: Natural Resource Management for Sustainable Agriculture and Environment, pp. 403-416, (ed.) DL Deb, Angkor Publs (P) Ltd. New Delhi.

Kaushik, B.D., 1987. Laboratory methods for Blue-Green Alage. Pub. Assoc. Publishing Co. New Delhi. pp.1-171.

Kaushik, P. 2000. Introductory Microbiology. Emkay Publications, Post Box 9410, B-19 East Krishna Nagar, Swami Dayanand Marg, Delhi. pp VIII+1-46.

Kaushik, P. 2003. Haridra (Turmeric): Antibacterial potential. Chowkhamba Sanskrit Series Office. K 37/99 Gopal Mandir Lane, Varanasi. pp 1-123 Coloured Pl. 16.

Kaushik, P. 2004, 2008. Microbiology: Questions and Answers. S. Chand & Co Ltd., New Delhi. pp xii+276.

Kaushik, P. and Chauhan, A. 2008. *In vitro* antibacterial activity of laboratory grown culture of *Spirulina platensis*. *Ind. Journal of Microbiology*. Vol. 48. No. 3 : 348 - 352.

Kaushik, P. and Chauhan, A. 2008. Antibacterial potential of aqueous and organic extracts of *N. commune* a cyanobacterium. VEGETOS (An International Journal of Plant Sciences). Vol. 21, N0.1 : 77-80.

Kaushik, P. and Chauhan, A. 2008. Cyanobacteria, The Future Therapeuticals at National Conference on Biodiversity, Conservation and Sustainable Development. Nagpur, India. 7th- 9th January.

Kaushik, P. and Chauhan, A. 2007. Cyanobacterium *Nostoc commune* as a source of antibacterial substances at National Symposium on Recent Trends in Plant Sciences and Herbal Medicine. Faizabad, India. 17th & 18th December.

Kaushik, P. and Chauhan, A. 2006. "*In vitro* antimicrobial activity of different extracts of *Anabaena variabilis*" at National Conference on New Vistas in Microbiology and Plant Sciences, Haridwar, India. 19-20 September.

Kaushik, P., Chauhan, A., Garima and Goyal, P. 2008. Antibacterial activity and UV-HPLC analysis of laboratory-grown *Nostoc muscorum* communicated.

Kaushik, P., Chauhan, A., Garima and Goyal, P. 2008. Antibacterial potential and UV-HPLC analysis of *Anabaena variabilis*. communicated.

Kaushik, P., Goyal, P., and Surbhi. 2008. Impact of microbial diversity in restoration of forest ecosystem. In Trivedi, P.C. (Editor). *Palaeobotany to Modern Botany*. Pointer Publishers, 807 Vyas Building, Chaura Rasta, Jaipur, pp 94-99.

Kaushik, P. and Goyal, P. 2008. *In vitro* evaluation of *Datura innoxia* (thorn-apple) for potential antibacterial activity. *Indian Journal of Microbiology* (Accepted in Dec, 2007) DOI: 10.1007/s12088-008-0020-7.

Kaushik, P. and Goyal, P. 2008. Antibiotic Resistance: A Global Problem. In: Souvenir of "National Conference on Biodiversity, Conservation and Sustainable Development (NACBCSD-08)" pp 29-33.

Kellam, S.J., Cannell, R.J.P., Owsianka, A.M., and Walker, J.M. 1988. Results of large-scale screening programme to detect antifungal activity from marine and fresh water microalgae in laboratory culture. *Brit. Phycol J.* 23, 45-47.

Khaleafa, A.F., Kharboush, M.A.M., Metwall, A., Mohsen, A.F. and Serwi, A. 1975. Antibiotic action from extracts of some seaweeds. *Bot Mar.* 18, 163-165.

Khan, Z., Bhadouria, P., Bisen, P.S. 2005. Nutritional and therapeutic potential of *Spirulina. Curr. Pharm. Biotechnol.* 6(5), 373-9.

Kim, H. M., Lee, E. H., Cho, H.H., and Moon, Y.H.1998. inhibitory effect of mast cell-mediated immediate-type allergic reactions in rats by *Spirulina. Biochem. Pharmacol.* 55, 1071-1076.

Knubel, G., Larsen, L.K., Moore, R.E., Levine, I.A., and Patterson, G.M.L. 1990. Cytotoxic, antiviral indolocarbazoles from a blue green alga belonging to the *Nostocaceae. J. Antibiotics,* 43, 1236. 1239.

Kobayashi, A., Kajiyama, S.I., Inawaka, K., Kanzaki, H. and Kawazu, K.1994. Z Naturforsch C, 49,464.

Kobbia, I.A. and Zaki, D. 1976. Biological evaluation of algal filtrates. *Planta Med.* 30, 90-92.

Kodani, S., Ishida, K., Murakami, M. 1998. Dehydroradiosumin. A trypsin inhibitor from the cyanobacterium *Anabaena cylindrical. J. Nat. Prod.* 61, 854-856.

Koehn, F.E, Longley, R.E, Reed and J.K.1992. Microcolins A and B, new

Kreitlow, S., Mundt, S. and Lindequist, U. 1999. Cyanobacteria-a potential source of new biologically active substances. *Journal of Biotechnology.* 70, 61-63.

Kulik, M.M. 1995. The Potential for using Cyanobacteria (blue green algae) and algae in the biological control of plant pathogenic bacteria and fungi. *European Journal of Plant Pathology,* 101, 585-599.

Kushak, R. I., Drapeau, C., Van Cott, E. M. and Winter, H.H. 2000. Favorable effects of Blue-green algae *Aphanizomenon flos-aquae* on rat plasma lipids. JANA. 2, 59-65.

Lai, J.Y., Yu, J., Mekonnen B and Falch, J.R. 1996. Synthesis of curacin A, an antimitotic cyclopropane-thiazoline from a marine cyanobacterium *Lyngbya majuscula. Tetrahedron Lett.* 37, 7167-7170.

Larsen, L.K., Moore, R.E., Patterson, G.M.L. 1994. â-Carbolines from the blue green algae *Dichothrix baueriana. J. Nat. Prod.* 54, 419-421.

Lau, A.F., Siedlecki, J., Anleitner, J., Patterson, G.M., Caplan, F.R. and Moore, R.E. 1993. Inhibition of reverse transcriptase activity by extracts of cultured blue-green algae (cyanophyta). *Planta Med.,* 59, 148-151.

Lazaroff, N. 1973. Photomorphogenesis and nostocacean development. In Carr and whitton (Editors), The Biology of Blue-green algae. Blackwell, Oxford, and University of California Press, Berkeley, pp. 279-319.

Liu, X.J., Chen, F. 2003. Cell differentiation and colony alteration of *Nostoc flagelliforme,* an edible terrestrial cyanobacterium, in different liquid suspension culture. *Folia microbial.* 48, 619-626.

Loya, S., Bakhanashvili, M., Kashman, Y., and Hizi, A. 1995. Peyssonols A and B, two novel inhibitor of reverstranscriptase of HIV-1 by the natural sulfoglycolipids from cyanobacteria: Contribution of different moieties to their high potency. *J. Nat. Prod.* 61, 891-895

Loya, S., Reshef, V., Mizrrachi, E., Silberstein, C., Rachamim, Y., Carmcli, S., and Hizi, A. 1998. The inhibition of the reverse transcriptase of HIV-1 by the natural sulfoglycolipids from cyanobacteria: contribution of different moieties to their high potency. *J. Nat. Prod.* 61, 891-895.

Luesch, H., Yoshida, W.Y., Moore, R.E., Paul, V.J. and Mooberry, S.L. 2000. Isolation, Strucure Determination and Biological Activity of *Lyngbyabellin A* from the marine Cyanobacterium *Lyngbya majuscula. J. Nat. Prod.* 63, 611-615.

MacMillan, J.B., Ernst-Russell, M.A., De Roop, J.S. and Molinski, T.F. 2002. Lobocyclamides A-C, lipopeptides from a cryptic cyanobacterial mat containing *Lyngbya confervoides*. *J. Org. Chem.* 67, 8210-8215.

Madigan, M.T., Martinko JM., Parker J. 2000. Brock Biology of Microorganisms, eighth ed. Prentice Hall International, Inc., London pp-524-526.

Martine, Naessens and Canh Tran-Minh. 1998. Whole-cell biosensor for determination of volatile organic compounds in the form of aerosols Analytica Chimica Acta, Volume 364, Issues 1-3, Pages 153-158.

Mathew, B., Sankaranarayanan, R., Nair, P.P., Varghese, C., Somanathan, T., Amma, B. P., Amma, N.S., and Nair, M.K. 1995. Evaluation of chemoprevention of oral cancer with *Spirulina fusiformis*. 24, 197-202.

Matsuda, H., Okino, T., Murakami, M. and Yamaguchi, K. 1996. Aeruginosins 102-A and B, new thrombin inhibitors from the cyanobacterium *Microcystis viridis* (NIES-102). *Tetrahedron*. 52, 14501-14506.

Matusiak, K. and Krzywicka, A. 1975. Influence of the extracts of *Chlorella vulgaris* on the growth of fungi. *Acta Microbial Polon Ser*. B.7, 51-54.

McDonald, M.I., Harvey K.J., Sinclair, A. 1981. Antibacterial activity of hydrolyzed linseed oil and linolenic acid against methicillin-resistant *Staphylococcus aureus. Lancet,* 10-56.

Metting J. F.1996. Biodiversity and application of microalgae. *J. Ind. Microbial.* 17, 477-489.

Metting, B. and Pyne, J.W. 1986. Biologically active compounds from microalgae. *Enzyme and microbiological technology*. 8, 386-394.

Microcystis aeruginosa. Phytotherapy Res. 11, 93-96.

Mishima, T., Murata, J., Toyoshima, M., Fujii, H., Nakajima, M., Hayashi, T., Kato, T., and Saiki, I. 1998. Inhibition of tumour invasion and metastasis by calcium spirulan, a novel sulfated polysaccharide derived from blue-green alga, Spirulina platensis. *Clin. Exp. Metastasis*. 16, 541-550.

Moen, L. K., and Clark, G. F. 1993. A novel reverse transcriptase inhibitor from *Fucus vesiculosus*. Int. Conf. AIDS. 9,145-161.

Moikeha, S.N and Chu, G.W. 1971. Dermatitis-producing alfa *Lyngbya majuscula* Gomont in Hawaii. II. Biological properties of the toxic factor. *Journal of Phycology*. 7, 8-13.

Moon, R.E. and Martin, D.F.1981. Assay of diverse biological activities of material elaborated by a marine blue green alga, *Gamphosphaeria aponina.Microbios Lett.* 18, 103-110.

Moon, S.S., Chen, J.L., Moore, R.E. and Patterson, G.M.L. 1992. Calophycin-a fungicidal cyclic decapeptide from the terrestrial blue-green alga *Calothrix fusca*. *J. Org. Chem.* 57, 1097-1103.

Moore, B. S., Chen, J.L., Patterson, G. M.L. and Moore, R.E. 1990. Paracyclophanes from blue-green algae. *J. Am. Chem. Soc.* 112, 4061-4063.

Moore, R.E. 1996. Cyclic peptides and depsipeptides from cyanobacteria: a review. *J. ind. Microbiol.* 16, 134-143.

Moore, R.E., Carmeli, S.and Patterson, G.M.L. 1990. Tolytoxin and New scytophycins from three species of *Scytonema*. *J. Nat. Prod.* 53, N0-6, 1533-1542.

Moreau, J. Pesando, D., Bernad, P. *et al.* 1988. Seasonal variations in the production of antifungal substances by some Dictyotales (brown algae) from French Mediterranean coast. *Hydrobiology*. 162, 157-162.

Mtolera, M.S.P. and Semesi, A.K. 1996. Antimicrobial Activity of Extracts from six green algae from Tanzania. Current Trends in Marine Botanical Research in East African Region. 211-217.

Mundt, S., Kreitlow, S., Nowotny, A. and Effmert, U. 2001. Biological and pharmacological investigation of selected cyanobacteria. *Int. J. Hyg.Environ. Health.* 203, 327-334.

Murakami, M.K.S., Ishida, K., Matsuda, H. and Yamoguchi, K. 1997a. Micropeptin103, a chymotrypsin inhibitor from the cyanobacterium *Microcystis viridis* (NIES-102). *Tetrahedron lett.* 38, 3035-3038.

Murakami, M.K.S., Ishida, K., Matsuda, H. and Yamoguchi, K. 1997b. Microviridins elastase inhibitors from the cyanobacterium *Nostoc minutum* (NIES-26) *Photochemistry*. 45, 1197-1202.

Murakami, M.K.S., Ishida, K., Matsuda, H. and Yamoguchi, K. 1997c. Acyclicpeptide. anabaenopeptin B, from the cyanobacterium *Oscillatoria agardhii*. *Phytochemistry* 44, 449-452.

Nakashima, H., Kido, Y., Kobayashi, N., Motoki, Y., Neushul, M., and Yamamoto, N. 1987a. Antiretroviral activity in a marine red alga: Reverse transcriptase inhibition by an aqueous extracts of *Schizymenia pacifica*. *J. Cancer Res. Clin. Oncol.* 113, 413-416.

Namikoshi, M. and Rinehart, K.L. 1996. Bioactive compounds produced by cyanobacteria. *J. Ind. Microbio.* 17, 373-384.

NCCLS-National Committee for Clinical Laboratory Standards. 1997a. Methods for Dilution Antimicrobial Susceptibility Tests for Bacteria that Grow Aerobically, Approved Standard M7-A4, Wayne, Pa.

NCCLS-National Committee for Clinical Laboratory Standards. 1997b. Mehtods for Determining Bactericidal Activity of Antimicrobial Agents. Approved Standard M26-T, Wayne, Pa.

Nowotny, A., Mentel, R., Wegner, U., Mundt, S. and Lindequist, U.

Ogino, J., Moore, R.E., Patterson, G.M.L.and Smith, C.D.1996. Dendroamides, new cyclic hexapeptides from a blue-green alga. Multidrug-resistance reversing activity of dendroamide A. *J. Nat. Prod.* 59, 581-586.

Ohta, K., Mizushina, Y., Hirata, N., Takemura, M., Sugawara, F., Matsukage, A., Yoshida, S., and Sakaguchi, K. 1998.Sulfoquinovosyldiacylglycerol, KM043, a new potent inhibitor of eukaryotic DNA polymerases and HIV-reverse transcriptase type 1 from a marine red alga, *Gigartina tenella. Chem. Pharm. Bull.* 46, 684-686.

Ohta, S., Chang, T., Ikegami, N., Kondo, M. and Miyata, H. 1993. Antibiotic substance produced by a newly isolated marine microalga, *Chlorococcum* HS-101. *Bull. Environ. Contam. Toxicol.* 50, 171-178.

Ohta, S., Shiomo, Y., Kawashima, A., Aozasa, O., Teruyuki, N. 1995. Antibiotic effect of linolenic acid from the *Chlorococcus strain HS-101* and *Dunaliella primolecta* on methicillin-resistant *Staphylococcus aureus. J. Appl. Phycol.* 121-127.

Okino, T., QiS., Matsuda, H., Murakami, M. and Yamaguchi, K. 1997. Nostopeptins A and B, elastase inhibitors from the cyanobacterium *Nostoc minutum.* J. *Nat. Prod.* 60, 158-161.

Oranday, M.A., Verde, M.J., Lozano, S.J.M. and Waksman, N.H. 2004. Active fractions from four species of marine algae. *International Journal of Experimental Botany.* 165-170.

Oufdou, K., Mezrioui, N., Oudra, B., Loudiki, M., Barakate, M. and Sbiyyaa, B. 2001. Bioactive compounds from *Pseudanabaena* species (Cyanobacteria). *Microbios.* 106 Suppl. 1, 21-9.

Ozdemir, G., Karabay, N.U., Dalay, M.C. and Pzarbasi, B. 2004. Antibacterial activity of volatile component and Various Extracts of *Spirulina platensis. Phytother. Res.* 18, 754-757.

Pande, B.N. and Gupta, A.B. 1977. Antibiotic properties in *Chlorococcum humicolum* (Chlorophyceae). *Phycologia.* 16, 439-441.

Papendrof, O., Koing, G.M. and Wright, A.D.1998. Hirridin B and 2, 4-dimethoxy-6-heptadecylphenol, secondary metabolites from the cyanobacterium *Phormidium ectocarpi* with antiplasmodial activity. *Phytochemistry.* 49, 2383-2386.

Papke, U., Gross, E.M. and Francke, W. 1997. Isolation, Identification and determination of the absolute configuration of fischerellin B. A new algicide from the freshwater cyanobacterium *Fischerella muscicola. Tetrahedron Lett.* 38, 379-382.

Patterson, G.M.L., Baker, K.K., Baldwin, C.L., Bolis, C.M., Caplan, F.R., Larsen, L.K., Levine, I.A., Moore, R.E., Nelson, C.S., Teschappat, K.D., and Tuang,

G.D. 1993. Antiviral activity of cultured blue-green algae (Cyanophyta). *J. Phycol.* 29, 125-130.

Patterson, G.M.L., Baldwin, C.L., Bolis, C.M., Caplan, F.R., Karuso, H, Larsen, L.K., Levine, I.A., Moore, R.E., Nelson, C.S., Tschappat, K.D., Tuang, G.D., Furusawa, E., Furusawa, S., Norton, T.R. and Raybourne, R.B., 1991. Antineoplastic activity of cultured Blue-green algae (Cyanophyta). *J. Phycol.* 27, 530-536.

Patterson, G.M.L., Bolis, C.M. 1997. Fungal cell wall polysaccharides elicit an antifungal secondary metabolite (phytoalexin) in the Cyanobacterium *Scytonema ocellatum. J. Phycol.* 33, 54-60.

Patterson, G.M.L., Larsen, L.K., Moore, R.E., 1994. Bioactive natural product from blue-green algae. *J. Appl. Phycol.* 6, 151-157.

Pelczar, M.J., Chan, E.C.S and Krieg, N.R. 1993. Microbiology. Tata McGraw-Hill Publishing Company Ltd. New Delhi. pp-VIII, 1-918.

Pelegrin, Y.F. and Morales, J.L. 2004. Antibacterial activity in marine algae from the coast of Yucatan, *Mexico. Botanica Marina,* 47, 140-146.

Perez, C., Pauli, M. and Bazerque, P. 1990. An antibiotic assay by agar-well diffusion method. *Acta Biologiae et Medecine Experimentaalis,* 15,113-115.

Piccardi, R., Frosini, A., Tredici, M.R., and Margheri, M.C., 2000. Bioactivity in free- living and symbiotic cyanobacteria of the genus *Nostoc. Journal of Applied Phycology,* 12, 543-547.

Pietra, F. 1990. A Secret World: A Natural Products of Marine Life. Ist ed. Birkhauser, Basel.

Podschun, R., Ullman, U. 1998. "*Klebsiella* spp. as Nosocomial

Pathogens: Epidemiology, Taxonomy, Typing Methods, and Pathogenicity Factors". *Clinical Microbiology Reviews* 11, (4), 589-603.

Postgate, J. 1998. Nitrogen fixation, 3rd ed.. Cambridge University Press.

Pratt, R., Daniels, T.C., Eiler, J.J., Gunnison, J.B., Kumler, W.D., Oneto, J.F., Strait, L.A., Spoehr, H.A., Hardin, G, J. Milner, H.W., Smith, J.H.C., and Strain, H.H. 1944. *Chlorellin,* An antibacterial substance from *Chlorella. Science.* 99, No-2574, 351-352.

Pratt, R., Mautner, H., Gardner, G.M., Sha, Y.H., and Dufrenoy, J. 1951. Report on antibiotic activity of seaweed extracts. *J. Amer. Pharm. Assoc.* 40, 575-579.

Qureshi, M.A., Garlich, J. D., and Kidd, M.T. 1996. Dietary Spirulina platensis enhances humoral and cell-mediated immune function in chickens. *Immunopharmacol Immunotoxicol.* 18, 465-476.

Ramamurthy, V.D. 1970. Antibacterial activity of the marine blue-green alga *Trichodesmium erythraeum* in the gastro-intestinal contents of the sea gull *Laurus brunicephallus*. *Marine Biol.* 6, 74-76.

Rao, M., Malhotra, S., Fatma, T. and Rattan, A. 2007. Antimycobacterial Activity from Cyanobacterial Extracts and Phytochemical Screening of Methanol Extract of *Hapalosiphon*. *Pharmaceutical Biology*. Vol. 45 No.2. pp. 88-93.

Reshef, V., Mizrachi, E., Maretzki, T., Silberstein, C., Loya, S., Hizi, A. and Carmeli, S. 1997. New acetylated sulfoglycolipids and digalactolipids and related known glycolipids from cyanobacteria with potential to inhibit the reverse transcriptase of HIV-1. *J. Nat. Prod.* 60, 1251-1260.

Rezanka, T. and Dembitsky, V.M. 2006. Metabolites produced by cyanobacteria belonging to several species of the family *Nostocaceae.Folia Microbiol.* 51(3) 159-182.

Rinehart, K.L., Namikoshi, M., Choi, B.W. 1994. Structure and biosynthesis of toxins from blue-green algae (Cyanobacteria). *J.Appl. Phycol.* 6, 159-176.

Rippka, R., 1988. Isolation and Purification of Cyanobacteria. *Methods in Enzymology*. 167, 157-172.

Romay, C., Ledon, N., and Genzalez, R. 1999. Phycocyanin extract reduces leukotriene B4 levels in arachidonic acid-induced mouse-earinflammation test. *J. Pharm. Pharmacol.* 51, 641-642.

Rosell, K.G. and Srivastava, L.M. 1987. Fatty acid as antimicrobial substances in brown algae. *Hydrobiologia.* 151/152, 471-475.

Ryan K. J., Ray, C. G. 2004. Sherris Medical Microbiology, 4th ed., McGraw Hill. ISBN 0-8385-8529-9.

Safonova, E. and Reisser, W. 2005. Growth promoting and inhibiting effects of extracellular substances of soil microalgae and cyanobacteria on *Escherichia coli* and *Micrococcus luteus*. *Phycological Research,* 53, 189-193.

Sano, T.and Kaya, K. 1996. Oscillapeptin G, a tryrosinase inhibitor from toxic *Oscillatoria agardhii* . *J. Nat. Prod.*59, 90-92.

Sathiyamoorthy1, P., and Shanmugasundaram, S.1996.Preparation of cyanobacterial peptide toxin as a biopesticide against cotton pests. *Journal Applied Microbiology* and *Biotechnology*. Volume 46, Numbers 5-6, 511-513.

Schaeffer, D.J. and Krylov, V.S. 2000. Anti-HIV Activity of Extarct and Compounds from Algae and Cyanobacteria. *Ecotoxicology* and *Environmental Safety* 45, 208-227.

Schopf, W., Packer, B. 1987. Early Archean (3.3-Billion to 3.5 Billion Year Old) Microfossils from Warrawoona Group, Australia. *Science,* 70-72.

Schwartz, J., Shklar, G., Reid, S., and Trickler, D. 1988. Prevention of experimental oral cancer by extracts of Spirulina-Dunaliella algae. *Nutr. Cancer.* 11, 127-134.

Schwartz, R.E., Hirsch, O.F., Sesin, D.F., Flor, J.E., Chartrain, M, Fromling, R.E., Harris, G.H., Salvatore, M.J. and Yudin, K. 1990. Pharmaceuticals from cultured algae. *J. Indus. Microbiol.* 5, 113-124.

Shin, H.J.Matsuda, H., Murakami, M.Yamaguchi, K.1997. Circinamaide, a novel papain inhibitor from the cyanobacterium *Anabaena circinalis* (NIES-41). *Tetrahedron.* 53, 5747-5754.

Shitanda, I., Takada, K., Sakai, Y., Tatsuma, T. 2005. Compact amperometric algal biosensors for the evaluation of water toxicity. *Analytica Chimica Acta,* Vol. 530, No. 2, 191-197.

Sieburth, J. M. N. 1965. In: Pollutions marines par les microorganisms et les produits petroliers. Commission internationale pour 1, expoloration scientifique de la Mar. Mediteranee, Paris. p217.

Sims, J.J., Donnell, M.S., Leary, J.V. and Lacy, G.H. 1975. Antimicrobial Agents from Marine Algae. *Antimicrobial Agents and Chemotherapy*.7, No-3, 320-321.

Singh, I.P., Milligan, K.E. and Gerwick, W.H. 1999. Tanikolide, a toxic and antifungal lactone from the marine Cyanobacterium *Lyngbya majuscula. J. Nat. Prod.* 62, 1333-1335.

Singh, S., Kate, B.N. and Banerjee, U.C. 2005. Bioactive Compounds from Cyanobacteria and Microalgae: An Overview. *Critical Reviews in Biotechnology*, 25-73-95.

Skulberg, O .M, Codd, G.A., Carmichael, W.W., 1984. Toxic blue-green algal blooms in Europe. A growing problem. *Ambio* 13, 244-247.

Smith, G.D. and Doan, N.T. 1999. Cyanobacterial metabolites with bioactivity against phytosynthesis in cyanobacteria, algae and higher plants. *J. Appl. Phyco.* 11, 337-344.

Smitka, T.A., Bonjouklian, R., Doolin, L., Jones, N.D., Deeter, J.B., Yoshida, W.Y., Prinsep, M.R., Moore, R.E. and Patterson, G.M.L. 1992. Ambiguine isonitriles, fungicidal hapalindole-type alkaloids from three genera of blue-green algae belonging to the Stigonemataceae. *J. Org. Chem.* 57, 857-861.

Soltani, N., Khavari-Nejad, R.A., Yazdi, M.T., Shokrevi, S., Fernandez-valiente, E. 2005. Screening of Soil Cyanobacteria for Antifungal and Antibacterial Activity. *Pharmaceutical Biology.* 43, 455-459.

Spoehr, H.A., Smith, J.H.C., Strain, H.H., Milner, H.W. and Hardin, G.J. 1949. *Carnegie Inst. Wash. Publ.* 58-61.

Stanier, G. 1988. Fine structure of cyanobacteria. *Methods in enzymology*, 167, 157-172.

Stanier, R.Y., Kunisawa, R., Mandel, M. and Cohen, B.G. 1971. Purification and Properties of Unicellular Blue-green algae (Order-chrococales). *Bact. Rev.* 35, 171-205.

Stensvik, Skulberg, O.M., Underdal, B. and Hormazabal, V. 1998. Antibacterial properties of extracts from selected planktonic freshwater cyanobacteria-a comparative study of bacterial bioassays. *The Society for Applied Microbiology*. 84, 1117-1124.

Stewart, J.B., Bornemann, V., Chen, J.L., Moore, R.E., Caplan, F.R., Karuso, H, Larsen, L.K. and Patterson, G.M.L. 1988. Cytotoxic, fungicidal nucleosides from blue-green algae belonging to the Scytonemataceae. *J. Antibiotics*, 41, 1048-1056.

Tariq, V.N. 1991. Antifungal activity in crude extracts of marine red algae. *Mycol. Res.* 95, 1433-1436.

Todar, K. 2007. Pathogenic *E. coli.* Online Textbook of Bacteriology. University of Wisconsin-Madison Department of Bacteriology.

Tokuda, H., Nishino, H., Shirahashi, H., Murakami, N., Nagatsu, A., and Sakakibara, J. 1996. Inhibition of 12-O-tetradecanooylphorbol-13-acetate promoted mouse skin papilloma by digalactosyl diacylglycerols from the fresh water cyanobacterium Phormidium tenue. *Cancer Lett.* 104, 91-95.

T.N. & E.L. Taylor. 1993. The Biology and Evolution of Fossil Plants. Prentice Hall, New Jersey.

Trimurtula, G., Ohtani, I., Patterson, G.M.L., Moore, R.E., Corbett, T.H., Valeriote, F.A. and Demchik, L.1994.*J. Am. Chem. Soc.*, 116, 4729.

Tuney, I., Cadirci, B.H., Unal, D. and Sukatar, A. 2006. Antimicrobial activities of the extracts of marine algae from the coast of Urla (Izmir, Turkey). *Turk. J. Biol.* 30, 171-175.

Vaucher, J.P. 1803. Histoire des conferves d'eau douce, contenant leurs different modes de reproduction, et la description de leurs principales especes. *J. Paschoud, Geneva*, pp.1-285.

Venkatachala, B.S., Shukla, M., Sharma, M., Naqvi, S.M., Srinivasan, R. and Udairaj, B. 1990. Archaean microbiota from the Donimalai formation, Dharwar Supergroud, India. *Precambrian Res.* 47, 27-34.

Waterbury, J.B. and Stanier, R.Y., 1981. Isolation and growth of Cyanobacteria from marine and hypersaline environments. Vol 1, Pp. 221-223. Sparinger Veralag, Berlin.

Weckesser, J. and Jurgen, U.J., 1988. Cell wall and external layers of cyanobacteria. *Mehtods in Enzymology*. 167, 173-178.

Welch, A.M. 1962. Preliminary survey of fungistatic properties of marine algae. *J. Bacteriol.* 83, 97-99.

Wen, M.J. and Weici, T. 1984. Screening for antimicrobial activities in marine algae from the Qingdao coast, China. *Hydrobiologia,* 116/117, 517-520.

Witvrouw, M., and De Clercq, E. 1997. Sulfated polysaccharides extracted from sea algae as potential antiviral drugs. *Gen. Pharmacol.* 29, 497-511.

Yang, H., Lee, E., and Kim, H. 1997. *Spirulina platensis* inhibits anaphylactic reaction. Life Sci. 61, 1237-1244.

Zulpa, G., Zaccaro, M.C., Boccazzi, F., Parada, J.L., and Storni, M. 2003. Bioactivity of intra and extracellular substances from Cyanobacteria and lactic acid bacteria on "wood blue stain" fungi. *Biological control,* 27, 345-34.

Chapter 10

GLOSSARY

Acetylated sulfoglycolipid - An antiviral agent produced from a cyanobacterium *Oscillatoria raoi.*

Acquired immunodeficiency syndrome (AIDS) - A disease of the immune system that is caused by infection with human immunodeficiency virus (HIV), commonly transmitted in blood and body secretions such as semen. People with AIDS are highly vulnerable to life-threatening infections.

Aerial - Growing in a habitat above the ground or water.

Aeruginosin 102A - A thrombin inhibitor isolated from a blue green algae *Microcystis viridis.*

Akinete - Resistant "resting cell of some cyanobacteria that differentiates from a vegetative cell by developing a thick outer wall, enlarging and accumulating cyanophycin, glycogen, lipid, and carotenoids.

Algaecides - Chemical agents that kill the growth of the algae.

Ambigol A - Halogenated aromatic compound isolated from *Fischerella ambigua* which inhibits the enzyme cyclooxygenase and HIV reverse transcriptase.

Ambiguine A Isonitrile - Fungicidal Hapalindol-type alkaloids isolated from *Westiellopsis prolifica.*

Anabaena - It is a genus of filamentous cyanobacteria in which, trichomes are made up of spherical cylindrical vegetative cells; heterocyst may be intercalary or terminal.

Anabaenopeptin B - A novel potent protease inhibitor isolated from a cyanobacterium *Oscillatoria agardhii.*

Anthraquinone - An antibacterial compound isolated from the cells of the cultured cyanobacterium *Nostoc commune.* The structure of this compound was determined by spectroscopic methods, mainly NMR, infrared and mass spectrometry.

Antibacterial - Drug that kills or inhibits the growth of bacteria.

Antibiotic - A drug used to treat infections caused by bacteria and other microorganisms. Originally, an antibiotic was a substance produced by one microorganism that selectively inhibits the growth of another. Synthetic antibiotics, usually chemically related to natural antibiotics, have since been produced that accomplish comparable tasks.

Antibiotic resistance - The state achieved when bacteria overcome the mechanism used by an antibiotic to kill or inhibit bacterial growth.

Antimalarial - Drug effective in the chemoprophylaxis of malaria.

Antimycotics - Chemical agents which either kill or inhibit the growth of fungi.

Antiviral - Compounds which more or less inhibit the replication of viruses in cells, tissues or organisms –either directly or by inducing interferon synthesis.

Apical cell - The cell located at the end of a filament or thallus.

Archaea - One of the three domains of living organisms characterized by the presence of isoprenoid glycerol diethers or diglycerol tetraether lipids in their membranes and archael rRNA. Three domains were proposed by Carl Woese and his co-workers.

Asexual reproduction - Reproduction involving the cells of a single parent.

Autocolony - A miniature version of the parent colony produced asexually by one of its cells; a coenobium autospore - A nonflagellate spore.

Axial - Chloroplast located in the center of a cell along its longitudinal axis.

Bacteria - One of the three taxonomical domains (Archaea, Bacteria and Eucarya) of living organisms proposed by Carl Woese and his co-workers on the basis of rRNA studies. Both the Archaea and Bacteria are procaryotes (lack true nucleus).

Bactericide - An agent that kills bacteria.

Bacteriocin - Antibiotic that is active against bacterial strains.

Bacterium - Individual, microscopic organism composed of a single cell which is devoid of membrane bound true nucleus rather have a chromosome spread in an irregularly shaped region called nucleoid or chromatin body. Many, but not all, bacteria cause disease.

Basal - Concerning the base of a thallus or filament.

Benthic - Inhabiting the bottom of an aquatic environment.

Benthic - Refers to the mud, sand etc. and the indigenous organisms, at the bottom of a lake, sea or other body of water.

Biflagellate - Cells with two flagella.

Bioautography - Method used for detecting a trace amount of a substance in a complex mixture, the given substance being an essential growth requirement for the test organism used. The components of the mixture are initially separated by chromatography. The presence of a given substance, in a given chromatographic band, is then determined by extracting the band with an appropriate solvent.

Biopesticide - Biopesticide is any material of natural origin derived from living organisms, such as bacteria, plant cells or animal cells that are used in pest control.

Biotechnology - The simplest definition of biotechnology is "applied biology." The application of biological knowledge and techniques to develop products. It may be further defined as the use of living organisms to make a product or run a process.

Bloom- An abundant growth of plankonic algae and cyanobacteria generally at the surface of a body of water.

Borophycin – Metabolite (containing a boron) isolated from marine strain of cyanobacteria *Nostoc linckia* and *Nostoc spongiaeforme vertenue.* It exhibits potent cytotoxicity against human epidermoid carcinoma and human colorectal adenocarcinoma cell lines.

Brackish - Slightly salty water-bodies.

Broad- spectrum antibiotic – An antibiotic which acts on a wide range of types of bacteria.

β-carbolines - An antiviral compound produced from *Dichaothrix baueriana* exhibited activity against Herpes simplex virus type 2.

Calcium spirulan - A novel sulphated polysaccharide, or an antiviral compound isolated from a marine blue green alga *Spirulina platensis.* This compound selectively inhibits the entry of enveloped virus into the cell.

Calophycin - Antifungal compound isolated from the culture of *Calothrix fusca,* this compound has been reported to be effective against many fungi.

Calothrix – A genus of filamentous cyanobacteria in which the mature trichomes taper from base to apex and are composed of disc shaped, isodiametric or cylindrical vegetative cells.

Calothrixin A - The compound isolated from *Calothrix* sp. that inhibit DNA replication in *Bacillus subtilis.*

Calyptra - Thickened or enlarged tip of a cyanobacterial filament.

Cancer - A malignant tumor of potentially unlimited growth that expands locally by invasion and can spread throughout the body in a process called metastasis.

Carcinogen - Any agent which can cause cancer and appear to exert their effect by interacting with the DNA of the target cells; many carcinogens are also mutagens.

Centrifugation - It is used for the sedimentation of fine particulate matter or molecules in a liquid medium, or for separating different types of particulate matter, or macromolecules, within a given suspension or solution.

Chlorellin - This is the first partly identified antimicrobial compound isolated from unicellular green algae, *Chlorella* that exhibited inhibitory activity against both Gram-negative and Gram-positive bacteria. Chlorellin is composed of peroxides of unsaturated fatty acids.

Chlorophyll a - Primary photosynthetic pigment in algae and higher plants.

Chlorophyll b - Photosynthetic pigment found in higher plants, green algae, and some euglenoids.

Chlorophyll c - Photosynthetic pigment found in chrysophytes, synurophytes, diatoms, tribophytes, dinoflagellates, cryptophytes, and brown algae; includes c1 and c2 components

Chlorophyta - A division of the algal members which are characterized by a combination of features *e.g.* they contained chlorophylls a and b, β-carotene and xanthophylls.

Chloroplast - Membrane-bound organelle involved in photosynthesis

Chromatography – The separation of molecules in a mixture by any technique which exploits the differing extents to which the different molecules get distributed between a stationery phase and a contiguous mobile phase.

Chromatophore - An organelle that contains pigments; *i.e.* a plastid.

Chrysophytes - Mixotrophic algae with golden-brown pigmentation.

Chytrids - Group of small, colorless fungi that may parasitize algal cells.

Ciprofloxacin - An antibiotic which inhibited the growth of the Gram-positive and Gram-negative bacteria.

Circinamide - It is the first papain inhibitor isolated from a cyanobacterium *Anabaena circinalis* and is an unusual peptide - like substance.

Coagulase - An enzyme that induces blood clotting; it is characteristically produced by pathogenic *Staphylococci.*

Colony Forming Units (CFU) - The number of microorganisms that can form colonies when cultured using spread plates or pour plates, an indication of the number of viable microorganisms in a sample.

Cross walls - Transverse cell walls.

Cryptophycins - Isolated from *Nostoc* sp. GSV 224 and has exhibited potent cytotoxicity against human tumor cell line. It shows good activity against a broad spectrum drug sensitive and drug resistant murine and human solid tumors.

Culture - Population of microorganisms cultivated in an artificial growth medium. A pure culture is grown from a single cell; a mixed culture consists of two or more microbial species or strains growing together.

Curacin A - A novel lipid compound isolated from *Lyngbya majuscula*, is a potent inhibitor of microtubule assembly with very low IC_{50} values against L1210 leukemia cells.

Cyanelles - The blue-green plastids of the glaucophytes that originally resulted from an endosymbiotic event.

Cyanobacteria (blue green algae)- Prokaryotic, photosynthetic microscopic organisms which differ from other bacteria in possessing chlorophyll a and in carrying out oxygenic photosynthesis.

Cyanophyceae- Members of blue green algae.

Cyanovirin N - An antiviral peptide isolated from an aqueous cellular extract of the cyanobacterium *Nostoc ellipsosporum*.

Cytokinesis - Division of the cytoplasm during cell division.

Dehydroradiosumin - A novel potent trypsin inhibitory dipeptide, which was isolated from the freshwater cyanobacterium *Anabaena cylindrica*.

Dendroamids - Isolated from cyanobacteria *Stigonema dendroideum* and exhibited a reversing multi-drug resistance activity.

Deoxyribonucleic acid (DNA) - The substance of heredity. DNA is a large molecule that carries the genetic information necessary for all cellular functions, including the building of proteins.

Desmid - A group of unicellular or filamentous green algae.

Detritus - Dead or decaying organic matter.

Diabetes - Abnormal condition characterized by a lack of insulin or a resistance to insulin. The excretion of excessive amounts of urine is an early symptom.

Diatoms - Photosynthetic microalgae with siliceous cell walls.

Dinoflagellates - Unicellular algae that are often covered in thick, armoured plates.

Dioceious - Organisms in which male and female (or + and -) gametes are produced on different individuals.

Diterpenoid - Compound isolated from the cells of the cultured cyanobacterium *Nostoc commune* by means of bioguided isolation and has exhibitted antibacterial activity.

Double false branching - Growth pattern that occurs when a filament breaks apart (often at a dead cell or heterocyst) and both pieces emerge through the mucilage sheath and continue to grow.

Embryophytes - The group consisting of bryophytes and vascular plants.

Endosymbiont - An organism that lives inside another, benefiting both individuals.

Epilithic - Growing on the surface of rock or stone.

Epiphytic - Growing on the surface of plants or other algae.

Escherichia coli - A bacterial species which inhabits the intestinal tract of man and other warm-blooded animals. Although it poses no threat to human health, its presence in drinking water does indicate the presence of other, more dangerous bacteria.

Euglenoids - Organisms that are unicellular or colonial, colourless or pigmented, and may move with an undulating, shape-changing motion.

Eutrophic - Waters rich in dissolved nutrients (especially nitrogen and phosphorus); leads to accelerated growth of algae and plants that depletes oxygen levels and reduces biodiversity.

Eutrophication - The enrichment of a habitat or environment with inorganic materials particularly nitrogen and phosphorus compounds, which encourage the growth of plants and algae. Eutrophication may occur as a result of self purification.

Eyespot - A light-sensitive, red-pigmented spot usually found in the cell anterior of some flagellated algal cells or reproductive cells; sometimes called a stigma.

False branch - A branch formed by slipping to one side of a section or loop of trichome through the sheath; a branch not formed by lateral division of a cell.

False branching - Single false branching occurs when a filament breaks apart (often at a dead cell or heterocyst) and one piece emerges through the mucilage sheath and continues to grow; see also double false branching.

Filament - A chain of cell together with an investing sheath.

Fischerellin A - The most active allelochemical compound of *Fischerella majuscula* that directed against other cyanobacteria and photoautotrophic organisms.

Fischerella - A genus of filamentous, thermophilic cyanobacteria in which some of the cells in a mature trichomes divide in more then one plane, resulting in a partly multiseriate trichome with lateral uniseriate branches.

Flagellum - A long, whiplike organelle that usually protrudes from a cell and is used for motility.

Fragmentation - Type of asexual reproduction where the parent thallus breaks into pieces to form new individuals; may be accidental or genetically programmed.

Fucoxanthin - A brown carotenoid pigment used in photosynthesis, found in the golden-brown and brown algae.

Fungicide- Chemical agents which either kill or inhibit the growth of fungi. They are used for the treatment or prevention of fungal disease of man, animals and plants.

Fungus - A member of a class of relatively primitive vegetative organisms, some of which can cause disease in humans.

Gas vacuole - Areas in cells that refracts light because of it gaseous nature, often appearing red; composed of numerous gas vesicles.

Gas vesicles - A cylindrical structure with conical ends composed of a single protein envelope. In cyanobacteria, cylindrical structures used to increase cell buoyancy.

Gene - A unit of genetic material (DNA) that carries the directions that a cell uses to perform a specific function, such as making a particular protein.

Geosmin- Produced by certain aquatic actinomycetes and cyanobacteria (*Anabaena, Lyngbya* and *Oscillatoria* spp.) and which imparts an earthy or musty taste and odour to the water.

Glaucophytes - A group of algae that contain blue-green plastids called cyanelles, sometimes called Glaucocystophytes.

Gloiosiphones A and B - An antimicrobial compounds isolated from the marine algae *Gloiosiphonia verticillaris*. Crude extract of these compounds were active against both Gram-negative and Gram-positive bacteria.

Glycolipids - An antiviral compound isolated from *Oscillatoria trichoides*.

Golden-brown algae - Algal groups with golden pigmentation; *i.e.* the Chrysophytes, Synurophytes, Haptophytes, and Diatoms.

Gram-positive- A group of bacteria that retain the purple stain while identifying a bacterium when stained by Gram's stain.

Grana - Stacks of thylakoids found in the discoid chloroplasts of land plants and some green algae.

Green algae - Morphologically diverse group of algae with green-pigmented chloroplasts.

Growth - In microbiology, an increase in both cell number and cellular constituents.

Halophilic - Inhabiting a salty environment.

Hapalindole - Moderately potent fungicide which is found in terrestrial blue green alga *Hapalosiphon*.

Hapalosiphon - A genus of filamentous cyanobacteria in which thallus caespitose, floccose, thin, aquatic, filaments free, not coalescing laterally, cells in one or two rows, sheath present, continuously branched; branches irregularly lateral, true, often arising only on one side of the filaments, false branches present, branches erect from the primary prostrate filaments, erect branches as broad as and similar to the main filament, heterocysts intercalary.

Haploid - Having a single set of chromosomes.

Hepatitis A - A usually benign inflammation of the liver, caused by an RNA-containing virus that does not persist in the blood serum.

Hepatitis A is transmitted especially in food and water contaminated with fecal matter.

Hepatitis C - A disease that inflames the liver, caused by a single-stranded RNA-containing virus and usually transmitted by exposure to blood or blood products. Hepatitis C leads eventually to scarring in the liver and liver cancer.

Heptadecane - The volatile component isolated from *Spirulina platensis.*

Heterocyst- Is a specialized cell in some cyanobacteria that function in nitrogen fixation. It differentiates from vegetative cells by adding a new outer multilayered wall, reorganizing the thylakoid structure, and by developing polar granules adjacent to pores which connect to other cells in trichomes.

Heteropolar - Filaments with distinct basal and apical components or cells that are asymmetrical along the transverse axis.

Heterotrophic -Organisms that feed on organic matter from external sources, and are osmotrophic or phagotrophic.

Holdfast cells - Structure that attaches a thallus or filament to the substrate.

Hormocyst - Hormogonia like short chain of cells envelope or encapsulated by a thick or sheath.

Hormogonium - In filamentous cyanobacteria, short, usually motile segments that breaks off from the main filament and are used for asexual reproduction and dispersal.

Hormone - A substance which is formed in one part of a multicellular organisms and which, in small quantity, can bring about one or more specific response when translocated to other part of the same individual.

Hormothamnin A - A novel antimicrobial compound isolated form *Hormothamnion enteromorphoides*. This compound showed broad spectrum antifungal and antibacterial activity. The structure of hormothamnin A was determined to be a cyclicundecapeptide.

Human immunodeficiency virus (HIV) - A retrovirus that infects and destroys CD_4^+ helper T cells in the immune system, causing the marked reduction in their number and thus lowering resistance to life-threatening infections that characterize AIDS.

Hypnospore- A thick-walled resting cyst.

Hypnozygote - A thick-walled, nonmotile resting zygote that remains dormant for a period of time until germinating when conditions improve; found in green algae, dinoflagellates, and some other flagellated organisms.

Immunosuppressive - Complete or partial suppression of normal immune response.

In vitro - "in glass"; a biological or biochemical process made to occur outside a living organism. Traditionally this refers to experiments performed in a test tube.

In vivo - "in life"; a biological or biochemical process that occurs or is observed while occurring within the body of living organisms.

Indolocarbozole- An antiviral compound isolated from *Nostoc sphaericum* which exhibited activity against Herpes simplex virus type 2.

Infection - The growth of a parasitic organism within the body. (A parasitic organism is one that lives on or in another organism and draws its nourishment there from). A person with an infection has another organism (*i.e.* germ) growing within him, drawing its nourishment from the person.

Inoculum - Material used to introduce a microorganism into a suitable situation for growth.

Intercalary - Located in the middle of a thallus, not on the end.

Isogamy - Sexual reproduction involving two types of gametes that function differently but are morphologically identical.

Isopolar - Symmetrical along the transverse axis, having the same morphology at both ends.

Isthmus - In desmids, the constricted region between the two semi-cells that usually contains the nucleus.

Kawaguchipeptin –a bactericide compound produced from *Microcystis aerugionosa*.

Klebsiella pneumoniae - Gram-negative bacillus is a member of the family Enterobacteriaceae.

Laboratory - A place for doing tests and research procedures and preparing chemicals, etc. Although "laboratory" looks very like the Latin "laboratorium" (a place to labor, a work place), the word "laboratory" came from the Latin "elaborare" (to work out, as a problem, and with great pains), as evidenced by the Old English spelling "elaboratory" designating "a place where learned effort was applied to the solution of scientific problems."

Laxaphycins – Compound isolated from the cultured blue green algae *Anabaena laxa* which has been responsible for the antifungal and cytotoxic activity.

Linolenic acid - An antibacterial compounds isolated from *Limnothrix redekei* HUB051.

Littoral zone - Region near the shore of lakes, ponds, or large rivers lorica. A protective covering surrounding flagellated cells of some algal groups.

Lyngbyabrllin A- Cytotoxic compound which was isolated from marine cyanobacterium *Lyngbya majuscula.*

Macrolide - Macrolides are typically bacteriostatic and are active mainly against Gram-positive and Gram-negative bacteria. Macrolides inhibit protein synthesis by binding to the 50S subunit of 70S ribosome.

Malyngolide - an antibiotic isolated from the dichloromethane extract of a shallow water variety of the blue green algae *Lyngbya majuscula.* It is effective against *Mycobacterium smegmatis* and *Streptococcus pyogenes.*

Meristem - Portion of a thallus where the cells are able to divide and produce new growth.

Metaphytic - Growing among the floating microscopic community in the littoral zone that is often associated with aquatic plants.

Microbiology - Study of microorganisms.

Microcolin A and B - A immunosuppressive lipopeptdes isolated from a Venezuelan sample of the blue green algae *Lyngbya majuscula.*

Microginin 299 - A Compound isolated from *Microcystis aerugionosa* that act as a leucin aminopeptidase inhibitor.

Microorganism - Minute organisms such as bacteria, viruses, fungi and protozoa. Some microorganisms can cause disease in humans.

Micropeptin 103 – A compound isolated from *Microcystis viridis* which act as a chymotrypsin inhibitor.

Micropeptin 478-A - A plasmin inhibitor produced from a blue green algae *Microcystis aerugionosa*.

Microviridin- The compound isolated from cyanobacteria *Nostoc minutum* and exhibited potent clastase inhibitor.

Minimum bactericidal concentrations (MBC) - The lowest concentration of an antimicrobial that will prevent the growth of an organism after subculture on to antibiotic free media.

Minimum inhibitory concentrations (MIC) - The lowest concentration of an antimicrobial that will inhibit the visible growth of a microorganism after overnight incubation.

Mixotrophic - Cells with both photosynthetic and heterotrophic or phagotrophic nutrition.

Molecule - The smallest amount of a specific chemical substance that can exist alone. To break a molecule down into its constituent atoms is to change its character.

Monoecious - Organisms in which male and female (or + and -) gametes are produced on the same individual.

Monophyletic - A group of organisms that share a single common ancestor.

MRSA - Short for methicillin-resistant *Staphylococcus aureus*; a notorious hospital and community-acquired pathogen that has the characteristics of multi-drug resistance, virulence and toxicity.

Multicellular - Composed of more than one cell.

Multinucleate - Having many nuclei.

Multiseriate - Arrange with more then one row of cells in trichome dividing in more than one plane.

Necridic cells - Dead cells in cyanobacteria that disrupt the main filament to produce false branching or hormogonia.

Nodulapeptin A – A protracted toxic isolated from a blue green algae *Nodularia spumigena*.

Nostoc - A genus of filamentous cyanobacteria in which the vegetative cells are spherical, ovoid or cylindrical. The heterocysts are intercalary under most circumstances. Nostoc is characterized by a confluent gel holding mans of trichomes together, which may be spherical or ovoid. The young trichome formed hormogonia have terminal heterocysts.

Nostodione A - An anti-mitotic compound isolated from a blue green algae *Nostoc commune.*

Nostofungicidine- An antifungal Lipopeptide isolated from the field grown terrestrial Blue green algae *Nostoc commune*. The structure of Nostofungicidine was elucidated by chemical degradation and extensive NMR measurements including DOF-COSY, HOHAHA, HMBC and ROESY techniques.

Nostopeptin A and B - Isolated from the freshwater cyanobacterium, *Nostc minutum* and exhibited clastase inhibitory activity.

Oligotrophic - Waters low in nutrients and organic production, typically rich in biodiversity.

Oogamy - Sexual reproduction involving the fusion of a small, flagellated male gamete with a larger, nonflagellated female gamete.

Oogonium - A female reproductive structure capable of producing one or more eggs.

Organisms – An individual living thing.

Oscillapeptin G - Tryrosinase inhibitor isolated from cyanobacteria *Oscillatoria agardhii.*

Oscillatoria - A genus of filamentous cyanobacteria in which the trichomes are straight, motile and composed of disc shaped cells. The cells may be much shorter in length than broad. Generally transeverse septum or cross-wall is thinner than the longitudinal wall.

Osmotrophic - Organisms that feed on dissolved organic molecules absorbed from the outside environment.

Parietal - Near the outer or peripheral surface of a cell or thallus; commonly used to describe chloroplasts situated along the cell wall.

Parsiguine - A novel antimicrobial substance from *Fischerella ambigue* having broad spectrum antifungal and antibacterial activity.

Pathogen - A specific causative agent of a disease, such as a bacterium or virus.

Pellicle - The surface covering of euglenoid cells composed of proteinaceous strips in a helical pattern.

Periphytic - Growing attached on the surfaces of plants, other algae, or other substrates.

Phagotrophic - Organisms that feed by ingesting particulate organic carbon or intact cells.

Phormedium - A genus of cyanobacteria of LPP group.

Photosynthetic - Organisms that use sunlight as an energy source to synthesize carbohydrates from carbon dioxide and water.

Phototaxis - Movement in response to light.

Phylogenetic - Concerning the relationships of evolutionary development.

Phytoplankton - Floating or swimming microscopic algae and cyanobacteria.

Placoderm desmids - Unicellular or pseudofilamentous desmids with cell walls made up of two parts of different ages separated by a central isthmus that are covered in pores.

Plankton - Microorganisms, small invertebrates and fish larvae etc. which drift more or less passively in water currents within the pelagic zone of lakes, seas and other bodies of water.

Plectonema - A genus of cyanobacteria of LPP group. Although, non-heterocysteous but can carry out nitrogen fixation under microaerobic conditions.

Polar nodule - In cyanobacteria, thickened regions where the heterocyst cell wall attaches to a vegetative cell.

Polyeders - In *Hydrodictyon* and *Pediastrum*, polygonal cells produced from zygote germination that give rise to new coenobia.

Polyketide- A compound which is, or may be regarded as a condensation product of acetate; polyketides. Metabolites derived from polyketides include many antibiotics and mycotoxins.

Polyphyletic - A group of organisms descended from more than one ancestor, may be more closely related to other organisms outside of the taxonomic group.

Prostaglandin - Fatty acid containing cyclopentane ring formed mainly by mast cell, monocytes and neutrophils and capable of promoting vasodilation-can exert anti-inflammatory effect.

Protoplasm - The living substance of a cell.

Pseudofilament - Loose chain of individual cells held together by mucilage or attached at the end walls, rather than sharing a common cell wall as in a true filament.

Pseudomonas - Genus of Gram-negative bacteria. They are rod-shaped and motile, possessing one or more polar flagella. Several species produce characteristic water-soluble fluorescent pigments. They are found in soil and water.

Pyrenoid - A proteinaceous structure associated with algal chloroplasts that often forms storage compounds.

Quadriflagellate - Cells with four flagella.

Rhizoid – Root-like extensions, usually used to attach a thallus to the substratum.

Rhodophyceae - Red algae are marine organisms; which occur in freshwater and soil habitats. In most red algae, the thallus is a branched filament or ribbon.

Salmonella - A group of bacteria that cause typhoid fever and a number of other illnesses, including food poisoning, gastroenteritis and enteric fever from contaminated food products.

Scyptolin – Cyclic desipeptides with elastase inhibiting activity from terrestrial cyanobacterium *Scytonema hofmanni*.

Scytonema - A genus of filamentous cyanobacteria in which the mature trichomes are uniform in width and are composed of disc shaped isodiametric or cylindrical vegetative cells.

Scytophycins A-E - Isolated from a cultured terrestrial blue green alga *Scytonema pseudohofmanni* and exhibit cytotoxic and fungicidal activity.

Sexual reproduction - Reproduction involving the cells of a single parent silica deposition vesicle involved in the formation of the siliceous valves of diatoms and siliceous scales of Chrysophytes and Synurophytes.

Spirulina - A genus of filamentous cyanobacteria in which trichomes are helical with little or no constriction between adjacent cells.

Staphylococcus - A genus of Gram-positive bacteria that includes pathogens, such as *S. aureus*, that most commonly infect the skin and mucous membranes.

Streptococcus - A genus of bacteria that includes important pathogens of humans such as those that cause streptococcal sore throat and infective endocarditis

Streptomyces - Genus of Gram-positive spore forming bacteria that grow slowly in soil or water as a branching filamentous mycelium similar to that of fungi. Important as the source of many antibiotics, for example streptomycin, tetracycline, chloramphenicol, macrolides.

Striae - Term used to describe elongated markings on the cell walls of desmids and diatoms.

Subaerial - Growing on or near the earth's surface.

Sulcus - In dinoflagellates, a longitudinal groove in the ventral side of the cell that holds the longitudinal flagellum.

Sulfolipid – It has been screen from *Phormidium tenue* which is known for its HIV-1 inhibitory effects in several T-cell lines.

Suture - In dinoflagellates, the regions between thecal plates.

Symplostatin 1 - Compound isolated from blue green algae *Symploca hydnoides* which showed cytostatic activity.

Syngamy - The fusion of two gametes during fertilization.

Synurophytes - Motile flagellates covered by siliceous scales; the scaled chrysophytes.

Tanikolide – A new brine shrimp toxic and antifungal compound that has been isolated from the lipid extract of the marine cyanobacterium *Lyngbya majuscula*.

Tannins - A heterogeneous group of complex polyhydroxybenzoic acid derivatives which occur in plants and in brown algae.

Tenuecyclamides A-D- Isolated along with known antibiotic, borophycin from *Nostoc spongiaeforme* var. *tenue*. The planar structure of Tenuecyclamides A-D was determined by homonuclear and inverse –heteronuclear 2D-NMR techniques as well as by high resolution mass spectrometry measurements.

Terminal - Located at the end of a thallus.

Thallus - General form of an alga that, unlike a plant, is not differentiated into stems, roots, or leaves; composed of many cells or filament. The thallus may have a definite shape but lacks internal differentiation.

Thylakoid - The intracellular membrane system that includes the reaction center, chlorophyll a and some carotenoid pigments.

Tjipanazole D - Isolated from the lipophilic extracts of *Fischerella ambigua* and is known to exhibit moderate antibacterial activity.

Tolybyssidins A and B - Compounds isolated from the culture medium of mass cultured cyanobacterium *Tolypothrix byssoidea* by means of bioguided isolation .The gross structure of these peptides were determined by 1D and 2D NMR and tandem mass spectrometry. The compounds exhibit moderate antifungal activity against the yeast *Candida albicans.*

Tolyporphin - A novel multi-drug resistance reversing agent isolated from the blue green alga *Tolypothrix nodosa.*

Tolypothrix - A group of filamentous cyanobacteria in which trichomes are uniform in with and composed of disc shaped.

Tolytoxin - A potent cytotoxic and fungicide, was first isolated from a terrestrial blue green alga *Tolypothrix conglutinata* var. *colorata.*

Toxins - Any of various microbial products or components which, when present at low concentrations in cell or tissues of a higher organism, can cause injury by interfering with the structural or functional integrity of those cells or tissues.

Trichogyne - In some charophycean green algae and in the red algae, the elongated portion of the female reproductive structure that receives the male gametes.

Trichome- A chain of cells without an investing sheath.

True branch – A branch formed by lateral or oblique division of a cell in trichomes.

Tuberculosis - A highly variable disease caused by the tubercle bacterium. Usually transmitted by inhalation of airborne bacteria.

It affects the lungs but may spread to other areas, especially the brain, from local lesions or by way of the lymph or blood vessels. Tuberculosis is characterized by fever, coughing, difficulty in breathing, and inflammation.

Turbidity - Measure of the cloudiness of a water-body due to suspended particles

Tychoplankton - Floating or free living organisms in shallow water of a lake intermingled with attached vegetation and periphyton, usually near shore.

Unicellular - Composed of a single cell.

Uniseriate - Arranged in or forming a single row or series of cells, such as in a branched or unbranched trichome.

Vegetative cells - Cells involved in growth, nutrition, or asexual reproduction, but not sexual reproduction.

Virus - Obligate intracellular parasite that must use a host cell's systems in order to replicate and proliferate. Some viruses cause infections in people.

Welwitinodolinones - An alkaloid isolated from a cyanobacterium *Westiella intricate.*

Westiellopsis - Thallus filamentous with true branching filament of two kinds, primary filament slightly thicker and more or less creeping, secondary filaments generally thinner and growing erect, filaments without a sheath and consisting of a single row of cells, heterocyst intercalary.

Xanthophyceae (yellow green algae)- A class of algae characterized by formation (during at least some stage of the life cycle) of motile biflagellated cells which have one long tinsel flagellum directed anteriorly, and one much shorter whiplash flagellum directed posteriorly. These occur mainly in freshwaters.

Zoospore- A flagellated spore.

Zygospore - A thick-walled resting spore formed from the fusion of gametes or a fertilized oogonium.

Zygote - The cell resulting from the fusion of two gametes.

INDEX

A

Acetylated sulfoglycolipid 173
Acutiphycins 13, 29
Aerial 173
Aeruginosin 16, 173
Akinete 46, 173
Algaecides 173
Amino acids 5
Anabaena 16, 17, 21, 29, 30, 33, 44, 45, 47, 48, 97, 99, 100, 108, 109, 110, 111, 127, 128, 137, 138, 139, 142, 145, 146, 147, 151, 152, 157, 162, 163, 169, 174, 177, 178, 181, 184
Anabaenopeptin 165, 174
Anthraquinone 33, 174
Antibacterial 1, 6, 11, 20, 22, 27, 28, 30, 31, 32, 33, 34, 35, 58, 92, 97, 108, 110, 112, 114, 116, 118, 120, 122, 124, 137, 138, 139, 140, 141, 142, 143, 145, 146, 148, 155, 156, 157, 158, 160, 161, 162, 164, 166, 167, 168, 170, 174, 179, 182, 184, 186, 190
Antibiotic 2, 3, 13, 22, 23, 27, 28, 30, 34, 77, 92, 94, 126, 138, 152, 153, 154, 155, 158, 160, 162, 166, 167, 168, 174, 175, 176, 177, 184, 185, 189
Antibiotic resistance 162, 174
Anticancer 15, 31, 157
Antihelmintic activity 15
Anti-HIV 6, 16, 17, 18, 19, 22, 141, 153, 154, 158, 169
Anti-inflammatory 14, 31, 188
Antimalarial 5, 174
Antimycobacterial 16, 44, 139, 142, 168
Antimycotics 5, 174
Antiviral 5, 6, 17, 18, 19, 22, 23, 30, 31, 142, 156, 158, 159, 163, 167, 171, 173, 174, 176, 178, 181, 183
Asexual reproduction 174, 180, 182, 191
Autocolony 175
Axial 175
Azolla 3

B

Bacteria 1, 2, 3, 4, 6, 7, 8, 13, 22, 23, 27, 28, 31, 32, 33, 34, 35, 45, 46, 52, 54, 58, 63, 67, 70, 72, 76, 97, 137, 139, 140, 141, 142, 143, 145, 146, 147, 148, 151, 156, 157, 163, 166, 171, 174, 175, 176, 177, 178, 179, 181, 184, 188, 189, 190
Bactericide 175, 183
Bacteriocin 157, 175
Basal 75, 175, 182
Benthic 3, 52, 175
Bg-11 medium 90
Biflagellate 175
Bioautography 175
Biofertilizer 12
Biomodulatory effect 14
Biopesticide 10, 11, 169, 175
Biosensor 11, 12, 164
Biotechnology 2, 153, 156, 158, 163, 169, 175
Bloom 23, 176
Borophycin 176, 189
Broad- spectrum antibiotic 176

C

Calcium spirulan 17, 18, 159, 164, 176
Calophycin 176
Calothrix 6, 17, 30, 33, 142, 156, 160, 165, 176
Calothrixin a 33, 156
Calyptra 52, 176
Cancer 21, 65, 154, 158, 164, 165, 169, 170, 176, 182
Carboxysomes 4
Carcinogen 176
Carotenoids 4, 14, 173
Catalase test 62
Chlorellin 22, 168, 177
Chlorophyll-a 4
Chlorophyll-b 4
Chlorophyll-c 4
Chlorophyta 177
Chloroplast 3, 141, 175, 177
Chromatography 31, 35, 175, 177
Chromatophore 177
Chrysophytes 177
Ciprofloxacin 92, 126, 127, 128, 129, 130, 132, 133, 135, 138, 143, 146, 177
Citrate utilization 77, 79
Coagulase 59, 61, 177
Colony 59, 62, 65, 68, 70, 74, 77, 91, 163, 175, 177
Colony forming units (CFU) 177
Comparative genomics 10
Cross walls 52, 178
Cryptophycins 178
Cultivation 4
Culture 178
Curacin 16, 163, 178
Cyanelles 178, 181
Cyanobacteria 1, 2, 3, 4, 5, 6, 7, 8, 9, 10, 11, 13, 14, 15, 16, 17, 18, 19, 20, 21, 22, 23, 28, 29, 31, 32, 33, 35, 36, 44, 45, 91, 93, 98, 141, 143, 149, 152, 153, 155, 156, 157, 158, 163, 164, 165, 168, 169, 170, 171,173, 174, 176, 178, 180, 181, 182, 185, 186, 187, 188, 189, 190
Cyanophyceae 178
Cyanovirin 178
Cylindrospermum 13
Cytokinesis 178

D

Dehydroradiosumin 163, 178
Dendroamids 178
Deoxyribonucleic acid 178

Diabetes 178
Diatoms 177, 179, 181, 188, 189
Dinoflagellates 177, 183, 189
Diterpenoid 32, 33, 161, 179
Double false branching 179, 180

E

Embryophytes 179
Enzyme inhibiting 6
Epilithic 179
Epiphytic 179
Escherichia coli 20, 27, 28, 29, 30, 31, 32, 33, 35, 44, 58, 67, 68, 69, 70, 97, 137, 151, 152, 157, 168, 179
Euglenoids 177, 179
Eutrophic 179
Eutrophication 5, 179

F

False branch 180
False branching 179, 180, 185
Filament 54, 174, 175, 176, 179, 180, 181, 182, 185, 188, 190, 191
Fischerella 13, 30, 31, 34, 35, 153, 156, 157, 158, 167
Fischerellin 158, 180
Flagellum 63, 68, 180, 189, 191
Fucoxanthin 180
Fungicide 180, 181, 190
Fungus 1, 29, 35, 141, 180

G

Gas vesicles 4, 180
Gene 180
Geosmin 181
Glaucophytes 178, 181
Gloeocapsomorpha 7
Gloiosiphones A and B 30, 155, 181
Glycolipids 6, 16, 168, 181
Golden-brown algae 181
Gram-positive 4, 6, 22, 23, 27, 28, 30, 31, 32, 33, 34, 35, 58, 63, 67, 140, 141, 142, 143, 146, 147, 148, 151, 177, 181, 184, 189
Green algae 1, 4, 17, 22, 27, 32, 45, 153, 156, 156, 157, 158, 159, 161, 163, 165, 167, 168, 170, 173, 177, 178, 181, 183, 184, 185, 186, 189, 190, 191
Growth 3, 4, 12, 13, 15, 23, 27, 29, 30, 31, 33, 35, 47, 48, 49, 50, 51, 53, 55, 56, 57, 60, 64, 66, 69, 73, 78, 91, 95, 97, 127, 128, 129, 130, 131, 132, 133, 134, 135, 137, 138, 141, 174, 175, 176, 178, 179, 180, 181, 183, 184, 185, 191

H

Halophilic 181
Hapalindole 30, 44, 156, 170, 181
Hapalosiphon 16, 30, 44, 45, 54, 57, 97, 107, 124, 125, 134, 135, 142, 145, 148, 151, 152, 168, 181
Haploid 181
Hepatitis a 181, 182
Hepatitis c 182
Heptadecane 34, 182
Herbicides 5
Heterocyst 54, 174, 179, 180, 182, 187, 191
Heteropolar 182
Heterotrophic 33, 182, 185
Holdfast cells 182
Hormocyst 182
Hormogonium 182
Hormone 182
Hormothamnin a 29, 182
Hypnospore 183
Hypnozygote 183

I

Immunosuppressive 5, 6, 22, 160, 183, 184
In vitro 14, 15, 17, 18, 36, 44, 155, 157, 158, 159, 162, 183
Indole reaction 70
Indolocarbazoles 13, 163
Infection 3, 65, 74, 77, 173, 183
Inoculum 91, 92, 94, 183
Intercalary 46, 54
Isogamy 183
Isopolar 183
Isthmus 183, 187

K

Kawaguchipeptin 160, 183
Klebsiella pneumoniae 20, 33, 58, 76, 77, 78, 97, 143, 145, 146, 151

L

Laboratory 3, 6, 23, 45, 68, 94, 97, 145, 146, 160, 161, 162, 166, 184
Laxaphycins 16, 30, 184
Lipid 15, 28, 32, 34, 35, 153, 178, 189
Lipopeptides 5, 164
Littoral zone 184
Lyngbya 15, 16, 18, 27, 44, 155, 158, 160, 163, 164, 169, 178, 181, 184, 189
Lyngbyabrllin A 184

M

Macrolide 184
Malyngolide 27, 155, 184
Metaphytic 184
Microbiology 63, 153, 156, 157, 158, 161, 167, 168, 169, 170, 181, 184
Microcolin A and B 184
Microfossils 8, 169
Microginin 299 184
Microorganism 174, 183, 184, 185
Micropeptin 103 185
Micropeptin 478-a 185
Microviridin 185
Minimum bactericidal concentrations 185
Minimum inhibitory concentrations 138, 139, 141, 146, 147, 185
Mirabilene 13, 29
Mitochondria 3
Mixotrophic 177, 185
Molecule 178, 185
Monoecious 185
Monophyletic 185
MRSA 185
Multicellular 3, 52, 182, 185
Multi-drug resistance 5, 74, 178, 185, 190
Multinucleate 185
Multiseriate 180, 185

N

Necridic cells 185
Nitrogen fixation 7, 13, 77, 167, 182, 187
Nodulapeptin a 185
Nostoc 1, 6, 13, 15, 16, 17, 18, 21, 29, 33, 35, 45, 46, 49, 50, 51, 52, 53, 97, 101, 102, 103, 104, 112, 113, 114, 115, 116, 117, 118, 119, 129, 130, 131, 132, 137, 139, 140, 142, 145, 147, 151, 152, 154, 156, 158, 160, 161, 162, 163, 165, 166, 167, 174, 176, 178, 179, 183, 185, 186, 189
Nostodione A 186
Nostofungicidine 161, 186

O

Oligotrophic 186
Organisms 2, 3, 4, 13, 21, 22, 33, 34, 52, 54, 68, 91, 92, 142, 146, 151, 174, 175, 178, 179, 180, 182, 183, 184, 185, 186, 187, 188, 191

Oscillapeptin g 168, 186
Oscillatoria 13, 16, 17, 18, 33, 44, 142, 160, 165, 168, 173, 174, 181, 186
Osmotrophic 182, 186
Oxidase test 65

P

Paracyclophanes 13, 29, 165
Parietal 186
Parsiguine 34, 158, 186
Pathogen 3, 27, 29, 63, 77, 185, 187
Pellicle 187
Periphytic 187
Phagotrophic 182, 185, 187
Phormedium 13, 16, 17, 18, 31, 141, 187
Photosynthetic 1, 4, 12, 13, 21, 177, 178, 179, 185, 187
Phototaxis 187
Phycobilisomes 4
Phycocyanin 4, 14, 15, 168
Phycoerythrin 3, 4
Phylogenetic 10, 22, 187
Phylogenetic data 10
Phytoalexin 167
Phytoplankton 161, 187
Placoderm desmids 187
Plankton 46, 187
Polar nodule 187
Polyeders 187
Polyketide 15, 187
Polyphyletic 187
Prostaglandin 188
Proteinase inhibitors 16
Protoplasm 188
Pseudofilament 188
Pseudomonas 20, 22, 23, 28, 29, 31, 32, 33, 35, 44, 58, 63, 64, 65, 67, 97, 137, 145, 151, 152, 188
Pyrenoid 188

Q

Quadriflagellate 188

R

Rhizoid 188
Rhodophyceae 30, 188

S

Salmonella 20, 27, 29, 30, 31, 33, 58, 72, 73, 74, 76, 97, 158
Scyptolin 188
Scytonema 188
Scytophycins 13, 29, 155, 160, 165, 188
Scytophycins a-e 188
Secondary metabolites 3, 10, 20, 21, 22, 24, 91, 97, 137, 146, 155, 158, 166
Sexual reproduction 183, 186, 188, 191
Spirulina 15, 16, 17, 18, 34, 44, 45, 52, 55, 97, 105, 120, 121, 131, 133, 151, 152, 153, 154, 159, 162, 164, 166, 168, 169, 171, 176, 182, 189
Staphylococcus 20, 22, 27, 28, 29, 30, 31, 32, 33, 34, 35, 44, 58, 59, 137, 145, 146, 151, 152, 164, 166, 185, 189
Sterilization 91
Stigonema 178
Streptococcus 189
Streptomyces 31, 189
Stromatolites 8
Subaerial 189
Sulcus 189
Sulfated polysaccharide 17, 19, 156, 159, 160, 164
Sulfoglycolipids 17, 18, 19, 141, 163, 164, 168
Sulfolipid 189
Suture 189
Symplostatin 1 158, 189

Synechocystis 35, 142
Syngamy 189
Synurophytes 177, 188, 189

T

Tanikolide 169, 189
Tannins 6, 22
Tantazoles 13, 29, 155
Tenuecyclamides a-d 154, 189
Terminal 46, 52, 63, 174, 186, 190
Thallus 46, 54, 174, 175, 180, 181, 182, 183, 184, 186, 188, 190, 191
Thylakoid 182, 190
Tjipanazole d 190
Tolybyssidins a and b 190
Tolyporphin 190
Tolypothrix 13, 14, 29, 154, 161, 190
Tolytoxin 13, 29, 155, 165
Toxins 2, 5, 6, 11, 22, 160, 168, 190
Toyocamycin 13, 14, 29
Trichogyne 190
Trichome 46, 49, 52, 180, 185, 186, 190, 191
Triple sugar iron 75, 86
True branch 190
Tubercidin 14, 29
Tuberculosis 44, 16, 139, 190, 191
Turbidity 91, 94, 191
Tychoplankton 191

U

Unicellular 11, 22, 170, 177, 178, 179, 187, 191
Uniseriate 180, 191

V

Vegetative cells 4, 46, 174, 176, 182, 185, 188, 191
Vibrio cholerae 33
Virus 17, 18, 52, 154, 159

W

Welwitinodolinones 191
Westiellopsis 30, 45, 54, 56, 97, 106, 122, 123, 134, 135, 137, 141, 142, 143, 151, 152, 174, 191

X

Xanthophyceae 191

Z

Zeaxanthin 4
Zone of inhibition 35, 92, 108, 109, 110, 112, 113, 114, 115, 116, 118, 119, 120, 121, 122, 123, 124, 125, 126, 146, 147, 148, 152

FSC
www.fsc.org
MIX
Papier | Fördert
gute Waldnutzung
FSC® C083411

Zeitfracht Medien GmbH
Ferdinand-Jühlke-Straße 7
99095 Erfurt, Deutschland
produktsicherheit@kolibri360.de